£18

0343041 - X

Fundamental tissue geometry
for biologists

Fundamental tissue geometry for biologists

K. J. Dormer

Reader in Botany, University of Manchester

CAMBRIDGE UNIVERSITY PRESS

CAMBRIDGE

LONDON   NEW YORK   NEW ROCHELLE

MELBOURNE   SYDNEY

Published by the Press Syndicate of the University of Cambridge
The Pitt Building, Trumpington Street, Cambridge CB2 1RP
32 East 57th Street, New York, NY 10022, USA
296 Beaconsfield Parade, Middle Park, Melbourne 3206, Australia

First published 1980

Printed in Great Britain at the University Press, Cambridge

ISBN 0 521 22326 1

Library of Congress Cataloguing in Publication Data
Dormer, K. J.
Fundamental tissue geometry for biologists
Includes bibliographical references and index
1. Tissues – Mathematical models   2. Geometry
I. Title
QL807.D67   574.8′2   79-14404
ISBN 0 521 22326 1

Contents

1

Introduction

The object of this book is to reduce to calculation the geometrical aspects of the growth and structure of living tissues. It has, therefore, an immediate practical application as an aid to routine computation in the histological laboratory. Such matters as the shapes and arrangements of cells, their relative volumes, and the shapes and areas of the interfaces between them, are to be expressed in formulae and equations, and by the tabulation of standardised numerical values, so far as the present state of knowledge will permit this to be done.

The geometry of uniform isotropic tissues such as mammalian adipose tissue or the simpler kinds of plant parenchyma is everywhere much the same (Fig. 1). It is not a simple geometry, and can be understood only by persons willing to follow a train of argument which is necessarily mathematical in form and of some little length. On the other hand it is not in any substantial respect a mysterious or unknown system. All its principal features can be derived quite naturally from equations which have perfectly clear and unambiguous biological meanings, and all its principal structural constants can be stated to respectable standards of numerical accuracy.

This generalised isotropic geometry constitutes a natural base level of histological organisation. Below this level it is rare for any living tissue to fall, from this level any more specialised type of differentiation must take its departure. The quantitative description of the most general form of tissue geometry is therefore of fundamental biological importance.

In relation to this serious scientific requirement the existing 'theoretical' literature of cell shape appears as little more than a historical curiosity. Even the best of the older published discussions are far too naive and elementary to effect any biologically fruitful analysis. It is also to be noted that of all the various ideas concerning tissue geometry which have circulated during the last half-century not one has been of a distinctively biological character, or originated by a biological author. For those wishing to enquire further, a sufficient number of key references will be given, but there is no reason why the general biological reader should be troubled with the citation as an 'authority' of every attempted importation of some rudimentary principle from the domains of mathematics or physics.

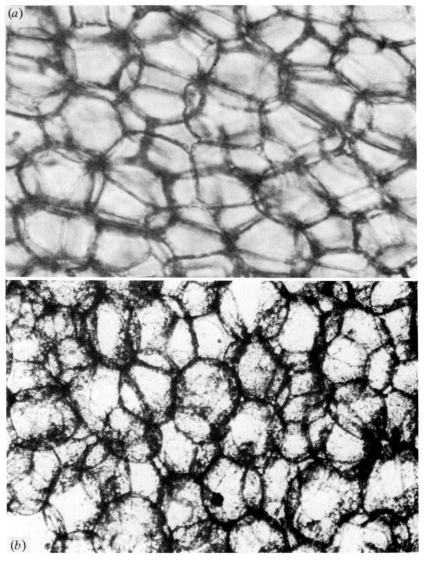

Fig. 1. Bird and apple, showing geometrical similarity of undifferentiated tissues.
(*a*) inner tissue from the shaft of a feather, (*b*) fruit flesh of a crab-apple. The apple
cells are nearly a thousand times the bulk of the bird cells (the magnifications here
being adjusted accordingly), their opaque contents give a different optical effect,
and their chemical composition is radically distinct from that of a feather.
Considered as geometrical patterns, however, the two tissues are virtually
interchangeable.

A book of this kind would be nothing without a large body of recorded observations upon which to test the validity of the theoretical reasoning employed. It is a pleasure to pay tribute to the devotion and accuracy of the workers whose contributions are acknowledged in Chapters 4 and 5. They have done their work so well that it would be impossible to make out any case for further repetition of the same kind of enquiry. Certain basic facts have been established for all time, and a stage of historical development has been effectively brought to a close. In the computation of some of the tables it has therefore seemed right to aim at completeness, and Tables 1, 3, 5 and 6 are believed to incorporate every published observation which is capable of being utilised for the purpose. Minor decimal inaccuracies and irregularities of coverage arise through technical defects such as the occasional misprint in the research journals, but it does not seem likely that the picture which is presented in these tables will undergo significant further change.

A small number of estimable observational studies have been excluded from consideration. In the main, these are cases where a worker has deliberately sought out tissues of unusually simple construction. It is an understandable response to a complex group of problems, but the diametrically opposite approach has been adopted in this book, with what measure of success it must be for the reader to judge. In so far as it may be possible to find a solution possessing some quality of mathematical generality, all the simplified special examples will then succumb automatically. If one believes, even mistakenly, that the citadel can be taken, one does not launch attacks upon outposts.

It is essential however to take a realistic view of the practical limits arising from the laws of statistical significance and the fatigue of the human observer. No scientific principle is more inescapable than that which relates the establishment of a result to a size of sample and so to an estimate of the hours which must be spent at the microscope. One cannot read far in the literature of cell shape without encountering suggestions which are at once destroyed by the touch of this grim arithmetic. Where the most resolute investigator has not exceeded a sample of 200 cells, a proposition which would be verifiable only from a sample of 10 000 is not a scientific contribution at all, but mere empty vapouring.

It is practicable to carry out a comprehensive geometrical analysis of simple tissues in which all the cells are of the same kind and the main source of complication is the mitotic division of cells. In these conditions we can readily understand not only the situation of a single cell but also the dynamic balance of the system as a whole. One might expect *a priori* that if a tissue differentiated a little further so that two morphological categories of cell became distinguishable within it, then the accompanying mathematical analysis would progress by some natural step to a slightly higher grade of difficulty. This is not what happens when the thing is tried; in the event the nature of the problem is suddenly and dramatically transformed, old methods reach the limit of their application, and new methods have to be devised.

It is in fact precisely at this point of change that the most exciting aspects of tissue geometry begin to reveal themselves. In place of mere conformity to the mass statistics of the tissue in which they were formed, the cells start to exhibit individual trends of behaviour. By proper methods of observation and calculation it is possible to achieve a separation of the three components which general biological experience would lead us to seek in such a context: that is to say, the changes in the shape of a cell during tissue differentiation can be attributed in their due proportions to heredity, to environmental influence, and to the morphological status of the cell concerned.

Biologically, the study of tissue geometry at this level is precisely and logically cognate with the direct microscopic observation of a free-living unicell. A shape-change in a cell which is immersed in solid tissue can generally be detected only by a process of calculation, whereas a comparable alteration in *Amoeba* would be visually obvious. This is not a difference of biological principle; it is merely an adjustment of laboratory procedures to the technical requirements of different specimens.

No author could prepare a work of this kind without becoming aware that a section of the biological community is profoundly averse to the development of a biologically effective science of tissue geometry. The desire that the application of mathematical reasoning to histological problems should go no further than the most superficial level of enquiry is widely apparent in the literature and has sometimes been indicated with astonishing frankness (Thomson & Hull, 1934). The decision whether to engage in a particular class of work is necessarily a personal one, and the position of a man who resolves to make his contribution purely as an observer must naturally be respected. But this does not justify wilful obscurantism: those who choose to abstain from algebra are not entitled to expect similar self-denial from the rest of us, nor should they express indecent incredulity if we claim to understand things which they do not.

As for the justification of the treatment which is offered here, there are inevitably many points of style and arrangement and historical judgement for which the author must bear the sole responsibility. But in relation to the main subject-matter there is no intention to rely upon anything but objective scientific tests. Mathematical relationships are public property, at the free disposal of anybody who has paper and pencil; their applicability to biological situations must stand or fall by the comparison of observation with theoretical prediction. The book is not meant to contain anything the reader could not check for himself, though in some instances a large amount of computational work would be required for this. Having regard to the standards of numerical accuracy which it is practicable to attain in biological investigations, the precision of numerical statement which can be found in many earlier publications is simply ludicrous (e.g. angles given to a second of arc, etc.). Except where it has been necessary to fix upon a number of decimal places for tabulating purposes, or a quantity is of such theoretical significance that it may become the starting-point for further computation, an attempt has here been made to avoid these spurious claims to accuracy. Some of the calculations are very rough indeed, but what has been roughly calculated has been at any rate partly understood, which is more than can be said of anything which has not been calculated at all.

At certain points, a modicum of active participation by the reader (completion of a geometrical construction, or perhaps the plotting of a graph from tabulated values) can hardly be avoided. There is nothing of a specially difficult or recondite nature, but the material will be unfamiliar to many, and the self-constructed diagram will often be a more potent aid to comprehension than anything which a stranger could supply. The text has been arranged progressively in the hope that the reader will be enabled to acquire not merely a passive understanding of geometrical relationships in tissues, but sufficient facility and confidence to embark upon the creative use of geometrical methods in his own laboratory.

Mathematical and physical principles of construction

The shapes and arrangements of cells are subject to the ordinary rules of mathematics, and where these rules are not clearly distinguished it is common for biologists to fall into the folly of making laborious observations to demonstrate some point which ought never to have been in doubt. For example it is a matter of strict geometrical proof (p. 37) that in any sufficiently extensive system of linked polygons, the average number of sides per cell must be six exactly. So when Hein (1930) reported observational values of 6.059, 6.036 etc. his 'research' was a scientific nullity. He had embarked, as others have done before and since, upon a system of measurement so designed as to measure nothing except its own inaccuracy.

It is therefore a major part of our task to ascertain the limits of abstract mathematical proof. So long as a system is mathematically determinate there is no room for any biological phenomenon to show itself. Biology, in the present context, begins specifically at the point where there are two or more mathematically admissible results, between which the organism must choose upon some basis other than that of geometrical necessity. Before we can study this situation we need to be able to identify it, and this requirement has the effect of dividing our enquiry, logically and philosophically, into three stages. We have:

(*a*) to recognise, and as soon as possible to eliminate from further discussion, aspects of tissue structure deriving from pure mathematical necessity;

(*b*) to estimate, once the available principles of mathematical causation have been exhausted, just how much remains undecided, and how much freedom of choice is consequently left to the cells;

(*c*) within the limits fixed by (*b*), to devise practical schemes of laboratory observation for extracting information about 'biological' decisions taken by cells in specified plant and animal tissues.

Of these three necessary stages not even the first has been at all clearly appreciated by most biological authors.

There has, however, been a much more general willingness to allow for the operation in tissues of certain physical principles, and in particular it has come to be widely understood that cell surfaces are under tension, and that the structure of tissues therefore bears a certain resemblance to that of bubbles and foams. Unfortunately, the adoption of this idea has been altogether too uncritical. From perceiving that surface-tension considerations are likely to be important, many published discussions move almost imperceptibly to a belief that cells *must* conform closely to the patterns which would be dictated by an all-powerful principle of surface minimisation.

All such beliefs are manifestly absurd, on thermodynamic grounds alone. Every relevant material system, animate or inanimate, contains sources of disorder and irregularity amply sufficient to outweigh the restorative action of surface tension in anything but the very simplest cases. It is a fact of observation that bubbles in an extensive foam do *not* have minimal surface area for their volume, and any sound theoretical appreciation must be such as to predict this result. This is not a question of energy levels but one of entropy: even if the viscosity were zero and the surface tension infinite, the cells in a tissue still could not take up the shapes which a pure 'minimum-surface' theory would prescribe for them. Ordinary tissues exhibit a general tendency to convexity of cell form, and a scarcity of re-entrant angles. When this has been said, the assistance one can derive from surface-tension concepts is almost at an end.

The geometrical environment of a cell

The environment of a cell which is immersed in a solid tissue consists essentially of its interfaces with other cells. During the formation of a tissue, every cell, and every interface, must be deemed to increase in size more or less continuously. We have, furthermore, in all ordinary cases to provide for the occurrence of mitosis, and if our treatment is to have any claim to generality it must be anticipated that it will be rare for two contiguous cells to be in the same phase of the mitotic cycle. Because the process of cell division implies a variation in cell volume by at least a two-to-one ratio, we can at once extract the following principles:

(*a*) A cell must in general be unsymmetrical because its neighbours will be in different mitotic phases and consequently of different volumes. More detailed analysis will show that all the faces of a cell must be different in area, and have different growth rates.

(*b*) A cell will typically display the greatest asymmetry at the commencement of mitotic interphase, for we must expect the mitotic partition to be substantially flat and equatorial. In so far as a mother-cell may approach the (symmetrical) form of a sphere, its daughter-cells can only approach the (unsymmetrical) form of a hemisphere. A progressive change of shape during interphase is, therefore, a necessary feature in the construction of any histological theory.

These arguments are amply sufficient to dispose of any 'one-shape' mathematical model which may be proposed for tissue structures. Many authors have chosen some ideal geometrical figure for the body of a cell, and then attempted to compare real cells with their geometrical concept. The biological value of such enterprises is negligible because the thinking involved in them is essentially static. Investigators who have proceeded in this way have been seeking a single shape, whereas the *absolute minimum requirement* for a biological interpretation of the simplest tissues is a complete cycle of shape-changes running synchronously with the mitotic cycle. In this book, static concepts are used only in the early stages, and a more dynamic treatment is substituted as soon as the necessary algebra can be developed.

The growth rate of a cell interface can only be a matter of compromise between the two cells concerned. If the cells are of the same morphological class, differing in little more than their mitotic phases, then the nature of that compromise will automatically be expressed in the equations of shape related to the mitotic cycle. A mathematical theory which could not deal with the growth of the wall between two similar cells would necessarily be incapable of describing tissues in any biologically effective manner at all. At this very basic level it is hardly worth while to enquire whether the two cells might be, in any sense, in conflict over the growth of the wall between them. Age-difference across a wall is in any case reversible (for the 'older' cell will divide, and be replaced by 'younger' cells).

Where two adjoining cells come to differ in some more permanent manner, their interface becomes in most cases quite obviously the site of a conflict of interest. We are then confronted inescapably with the need to measure two phenomena which may be compared respectively with the concepts of stress and strain in relation to the strength of materials. Each cell, by adopting a discordant pattern of growth, exerts a force upon its neighbour, and the neighbour yields to that force to an extent which is limited by its own ability to resist. In a complex biological situation, it will be inexpedient to use the actual language of simple experiments in elasticity. But although different words may be employed, the similarity of principle is extraordinarily close: in Chapter 7 we shall find it quite practicable, and indeed necessary, to study the way in which the resistance of a cell is progressively overcome as the geometrical circumstances are made more demanding.

Historical perspective

There are aspects of tissue geometry so obvious that they can hardly escape the attention of any person who seriously considers the question at all. The appreciation that cells are polyhedral figures came with the very first histological reports ever published, and a recognisable corpus of theoretical concepts was in circulation well before the middle of the 19th century (Macior, 1960). In a field where the *bona fide* rediscovery of matters already recorded has evidently been commonplace, the assignment of true priorities can serve no scientific purpose in itself. An objection must however be raised to the unscholarly habit of attributing ancient ideas to modern authors as though they were recent discoveries. Such attributions impede scientific progress. They are also in very questionable taste. For example Thompson (1915, 1961) was a modest and exceptionally well-read man, accustomed to the citation of early literary sources. Naturally, he made no attempt to claim personal credit for things which were in print before he was born, but ignorant commentators have persistently misrepresented his position. In defiance of all the biographical evidence (Dobell, 1949; Bonner, 1961) he has had fathered upon him a vast medley of notions for which he had no original responsibility whatsoever.

Of the contrasting lines of thought which have been displayed in published work we may first consider the attention which has been given to surface-tension phenomena, drawing inspiration from the researches of physicists upon the stability of bubbles and liquid films. The facility with which a film may be formed, and its ability (by contracting to the smallest surface area which its circumstances will permit), to solve a difficult mathematical problem within a second or two, make it an irresistibly attractive experimental subject. The relevance to cell forms was recognised very early, and is quite undeniable. We have, therefore, an extensive literature dealing with particular configurations resulting from the division of space by partitions of minimum area. The problems which arise are investigated mathematically, by solving equations for minimisation of area, or experimentally, by creating liquid films across wire frames, perspex boxes etc. A good modern summary is that of Schüepp (1966). Such studies constitute, of course, a perfectly legitimate area of innocent enjoyment for those whose tastes lie in the relevant aspects of physics and mathematics. Whether they can offer much which is of value to the practical histologist is another question altogether. For the present we may simply note that many features of surface-tension theory have been placed before the biological public repetitively, and over a long term of years, without ever leading to significant biological application. It has for example been stated many times that the angle at the corner of a cell 'ought' to be $109° 28' 26''$. Whatever else such a statement may lead to, it certainly does not suggest a realistic research assignment for a histological laboratory. The truth seems to be that beyond a very elementary level, and outside a range of specially simplified situations like those which may arise in the early divisions of a spore or zygote, pursuit of the surface-tension principle easily degenerates into a rather pedantic study which affords little satisfaction to the working biologist in his daily contact with specimens.

From the much looser and more general comparison of polyhedral cell shapes with such inanimate models as foam and honeycomb, and from early experiments on such systems as the swelling of peas in a bottle, one might have expected a significant development of biological theory to arise. In particular the 17th-century observation that cells in the pith of a plant stem are commonly disposed in hexagonal columns ought to have led at once to the correct conclusion that cells tend to be fourteen-faced solid figures, by an obvious relationship which we shall use at p. 61. Very strangely, until the modern period the majority of authors did not take this easy and natural step, preferring to believe that cells should be twelve-faced, and specifying in many cases the rhombic dodecahedron as the chosen model for the typical cell. For this view, which prevailed over a long period of years in defiance of observational records which ought to have brought about its instant dismissal, it does not appear that there was ever any justification of a kind which would now be admitted to a scientific argument. It was much more a question of mystical or philosophical preconceptions about the 'perfection' of regular geometrical figures.

The modern period may be said to have begun with the first clear and general recognition that the typical figure of a cell in solid tissue would have fourteen faces rather than twelve. By a singularly unfortunate historical accident, this development, which could have been accomplished at any time by simple reference to observational facts, came to be complicated and obscured by extraneous issues which many later authors have failed to disentangle.

Mainly as an exercise in pure mathematics, but with some reference to soap films and the minimisation of surface area, Kelvin (1887) specified a particular fourteen-faced solid figure in a publication which has no real relevance to biological problems and which seems to have escaped the notice of biologists for nearly thirty years. The work was then 'discovered' in an episode which seems to have been very incompletely recorded. The printed sources are not sufficient even to identify convincingly the persons who actually initiated the attempt to find biological application for the Kelvin model. Because the introduction into biology of any fourteen-faced figure would at that time have appeared as a dramatically successful correction of the dodecahedral error, the Kelvin treatment immediately took a central position as the main theme of a literature which grew rapidly from about the year 1917 onwards. This literature is of a very distinctive and somewhat uniform character. It is intensely repetitive, showing hardly any change of theoretical outlook over a period of fifty years, and much of it is unhappily not of sufficient quality to justify serious consideration.

The mere exposure of errors is a distasteful operation which should not be pressed beyond the limits of genuine necessity, but in order to clear the way for a biologically sound discussion it is desirable to indicate briefly some of the specific features which do not satisfy the normal criteria of scientific and mathematical acceptability.

In the first place, there has been widespread presentation of numerical results which have not been subjected to any of the standard tests of statistical significance, and much unprofitable discussion of matters upon which there is evidently no prospect that statistically admissible results could be obtained. As another aspect of the same deficiency, sampling procedures have varied between the questionable: cells 'chosen at random with the following reservations' (Wheeler, 1955), and the downright perverse.

Secondly, there is the objection that when the Kelvin figure proves to be incompatible with observations upon real cells the facts are not accepted. Instead various forms of verbal prevarication are employed to conceal the difficulty. Consider for example the words of Lewis (1943*a*): 'Notwithstanding the rarity of its occurrence the orthoid tetrakaidecahedron (i.e. the Kelvin figure)...may properly be regarded as the typical shape of cells in masses.'

Thirdly there are widespread and disturbing indications of an insecure grasp of the fundamentals of geometry. This may be illustrated by reference to the problem of the 'tetrahedral angle'. The projecting corner or vertex of a cell is normally the meeting-point of three faces, as at the corner of a cube, but it is possible to envisage as an alternative a vertex formed by convergence of four faces, as at the summit of a square pyramid. The biological literature abounds in reports of this phenomenon, assessments of its frequency, and so on. In the course of this work some very unsound notions are expressed. Thus Matzke (1939) uses the argument that 'the likelihood of four faces meeting in a point becomes greater as the area of the individual faces becomes reduced'. But this is merely a statement, under very thin disguise, of one of the most notorious fallacies in the whole field of mathematics. Representing the suggestion as a diagram, it amounts to a claim that the chance coincidence of two intersections within the length of a line becomes more probable as the line is made shorter. This is disproved by the methods introduced a century ago by Georg Cantor in work which now takes an honoured position in the standard histories of mathematics. In order to achieve a consistent and satisfying theoretical treatment of cell shapes it is necessary to view all cell vertices upon a uniform basis: a reader who has difficulty in appreciating the force of this requirement in purely mathematical terms is advised to consider the situation at a cell vertex upon a molecular or atomic scale of magnification. The concept of the four-faced vertex will then emerge in its true capacity, not as a report of objective reality in the specimen, but merely as a limitation of the technical possibilities in observational histology.

Lastly we may note with regret the occurrence of mathematical blunders of a far more elementary nature. Some of these concern stackability, the ability of a given solid figure to combine uniformly with others of its own kind to fill space completely without leaving gaps: the hexagonal prism is stackable, the sphere is not. The determination of stackability requires no more than mechanical accuracy in arithmetic; nevertheless Millis (1918, 1926) persistently denied the stackability of the Kelvin figure and seems to have recanted only in private (Gross, 1927, in footnote) while Lewis (1935) claimed to have a uniformly stackable figure with eighteen faces, which can only mean that he failed to check his working, as a uniformly stackable 18-hedron is a known structural impossibility (see p. 61).

Surveying the whole of the literature since 1917 it is impossible to overlook the pre-eminence of F. T. Lewis, whose works uniquely contain virtually all the biologically sound principles which can be found circulating among the workers in this field. In view of this situation, the major contributions of Lewis are listed in the bibliography, though not all of them are individually cited in the text. In relation to the arrangement of this book, the ideas drawn from the literature are those of Chapter 2 and those of that section of Chapter 3 which deals with polygons on closed surfaces, in terms mathematically related to Euler's theorem. Most of these concepts are expounded in the long series of papers by Lewis: textual analysis leaves no doubt that Lewis was without any power of origination in mathematics, and depended entirely on mathematical colleagues for the principles for which he sought biological exemplification, but his achievement was nevertheless an outstanding one. The subject has had better observers, but no theorist of comparable stature. There is circumstantial evidence that contact with Lewis may have provoked the mathematician Graustein into making his own curiously isolated contribution (Graustein, 1931).

We must also briefly notice the existence of a class of publication which considers the consequences of aggregating together in large numbers polyhedral figures of some chosen ideal form, whether the Kelvin 14-hedron or some other. Mathematically such studies relate to the classical textbook problems on the piling of cannon-balls etc., and it is normally a simple matter to develop formulae for the number of units required as successive layers are added to the assembly. Examples can be seen in Marvin (1936) and Naum & Matzke (1955), while Glaser & Child (1937) attempted to use an aggregation algebra as the basis of a growth-curve. These are all perfectly legitimate forms of geometrical construction but from the biological point of view they present a totally artificial appearance.

These exercises in the aggregation of polyhedra reveal in a particularly acute form some philosophical weaknesses which are apparent also in other sections of the biological literature. The attitudes adopted by Lewis and others are not truly analytical but merely contemplative. Where one would hope to see the active development of a powerful and creative algebra along distinctively biological lines one finds all too often simply a parade of examples and special cases. In its extreme forms this approach results in the pictorial representation of each individual cell in a sample of 50 or 100, apparently on the assumption that the emergence of general principles must follow automatically if only the array of diagrams can be long enough continued.

For a reader accustomed to the style of the more progressive sciences, the general impression is an unfavourable one. The biological literature relating to tissue structure appears in fact very largely to have failed through pusillanimity. There has not been the courage to try any hypothesis which could not be instantly 'verified' by direct comparison with a picture of a specimen. From this point of view the main innovation in this book is the willingness to resort to hypotheses which are *not* self-evidently true, but verifiable only by observational test of their slightly more distant consequences.

Choice of examples

This book is based on the axiom that there must exist a single universal set of structural parameters, such that the geometry of any living tissue, past, present or future, can be specified by giving those parameters the appropriate set of numerical values. If we compare a portion of human brain with a block of mahogany, all the geometrical contrasts must ultimately be reducible into a common standardised table of constants, containing two numerically different values for each entry. Unless such a reduction is possible there is no point in our proceeding further, because our studies will not qualify for recognition as a branch of science.

Except that it will be convenient to use accessible materials in ascending order of complexity, it can, therefore, never be a matter of consequence which specimen is chosen to exemplify a principle. The relevant research literature displays an enormous disproportion in favour of plant tissues, and it would be quite impracticable for any contemporary author to disguise this imbalance, even if there were any particular reason for wishing to do so. Botanists have probably had more incentive to concern themselves with details of cell shape, but it appears also that animal tissues are so troublesome in the laboratory that even zoologists tend to lose patience with them, and transfer their attention to vegetable products.

In seeking to establish a general theory, we have no need to concern ourselves with any situation which is capable of being treated as a simplification of normal tissue structure. When, for example, a tissue becomes partly or wholly coenocytic, the geometrical record of events is erased, and ceases to be available as an object of study. Some of the biologically important tissues of animals (notably muscle) are almost empty of geometrical interest in their mature state. Again when cells separate from one another this is unlikely to raise any geometrical consideration which will justify extended enquiry or discussion; free cell surfaces bordering on intercellular spaces are merely withdrawn from general involvement in the affairs of the tissue, and placed under a simplified regime approximating to that of isolated unicells. The physiological implications may be profound, but the geometrical ones are completely trivial.

It is necessary therefore to allow a geometrical treatise to adopt its own scale of values in the selection of examples. This is a book about the biological importance of tissue geometry, which is not at all the same thing as the geometry of biologically important tissues. As in other biological disciplines, specimens for geometrical analysis must be chosen largely on grounds of amenability to the particular laboratory techniques which it is desired to practise upon them.

2

Putting cells together

A cell is a soft and deformable body, with its surface in tension and its centre in compression, so that the profile tends to be convex. These at least are the typical conditions from which our analysis ought to begin. If two cells are to meet upon an interface, both cells must undergo a degree of local flattening, and this is a process which almost demands to be demonstrated by the use of models. We have accordingly many records of trials in which peas have been poured dry into a bottle and then caused to swell by addition of water, lead shot have been compressed in steel cylinders, balls of clay have been forced together, bubbles have been stacked in beakers, and so on. At a moderately casual level of inspection all such experiments yield the same general result, and one which is entirely consistent with general impressions of many plant and animal tissues. What is produced is an array of polyhedra, facetted solid figures of which a cut diamond affords perhaps the readiest exemplification from ordinary life. Polyhedra formed by the compression of roughly equal spheres will for the most part remain pretty symmetrical. A substantially globular general figure will persist in a majority of cases. The surface area of such a polyhedron is greater than that of the parent sphere, but the increase in surface will not be very large. A polyhedron which is not markedly lopsided can approximate more closely to a sphere as the number of its own faces is larger, and we may so far anticipate as to say that cells on average are 14-hedra or thereabouts. We shall presently discover that the projecting points or vertices of a 14-hedral cell are exactly 24 in number, which means that each of them can be much blunter and less prominent, for example, than the vertices of a cube, of which there are only 8. It is of course quite possible to design complicated polyhedra with sharp projecting points, but such figures are rare in biologically unspecialised tissues, and they do not occur at all in compression models which start from the packing of spheres.

The pressing of free cells into contact is a biologically possible mode of growth but an uncommon one. Most cell shapes are generated in a different manner, by repeated mitosis in tissues where no cell except the first can ever conceivably have been spherical. Despite this stark contrast of developmental history, the similarity of appearance between tissues and compression models is extraordinarily close. A model which contains no counterpart of the mitotic process hardly seems likely to render a satisfactory account of natural cell-shapes, but in fact a fairly critical standard of analysis is needed to reveal the expected deficiency.

Curvatures and pressures

The common feature of all compression models is an equilibrium between tensile forces in the outer shell of each unit and compressive ones in its central core. This has not always been understood. For example Marvin (1939a) introduced his lead shot model as 'a physical system in which surface tension is reduced to a minimum'. It seems incredible that anybody should claim the tensile strength of any solid metal to be a 'minimum' (whatever that may mean): of course if the surface material were not to be in tension it would necessarily rupture, which in Marvin's own experiments it seems it did not. But whereas in the deformation of solids it is extremely difficult to determine the exact depth at which surface tension gives place to central compression, a bubble affords complete separation of tension in the liquid phase from compression in the gas. From this simplification we need not hesitate to draw every available advantage. Much of what follows may be understood in terms of bubble models. Use may also be made of bubble half-models, hemispheres created on a bath of liquid and brought together in combinations as required.

Consider a sphere of radius r in which a pressure p is contained by a tension t. To obtain an equation, dissect the sphere into two hemispheres (Fig. 2) and balance the compressive force tending to separate the hemispheres with the tensile force holding them together. The equilibrium is:

$$\pi r^2 p = 2\pi r t,$$

so that

$$p = \frac{2t}{r}.$$

For bubbles, t is constant, not changing as the film expands and contracts. This is different from the behaviour of an inanimate solid membrane such as a rubber sheet, which becomes taut when it is stretched and slacker as it is released. As between these two extremes, the position of cellular surfaces is intermediate, complicated, and very much a matter of debate. Let us *provisionally* accept the restriction that t is to be constant, and exploit the fact that curvature and pressure then stand in a simple reciprocal relationship.

We shall not be able to limit our discussion to spherical surfaces and may as well proceed at once to the higher level of mathematical generalisation which will be needed. In Fig. 3 an open-ended cylindrical

Putting cells together

$\pi r^2 p$

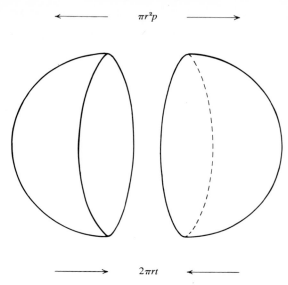

$2\pi rt$

Fig. 2. Balance of forces in a spherical skin stretched by
internal pressure.

$2rlp$

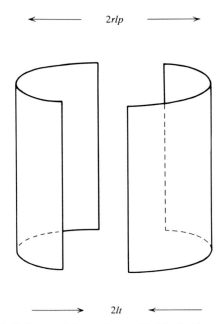

$2lt$

Fig. 3. Balance of forces in a cylindrical skin stretched
by internal pressure.

surface of length l has been dissected into two semicylinders giving the equilibrium:

$$2rlp = 2lt,$$

so that

$$p = \frac{t}{r}.$$

This is just half the pressure deduced above for a spherical surface, and the difference reflects the fact that a cylinder has curvature in only one

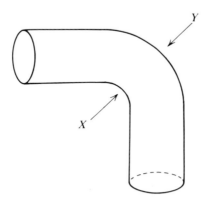

Fig. 4. Sign convention for curvature of surfaces, illustrated by a pipe-bend. At X the surface of the pipe is externally concave round the bend but externally convex round the bore of the pipe. In calculating the balance of forces these curvatures must be given opposite signs. At Y both kinds of curvature have the same sign.

direction (circumferentially but not along its axis) whereas a sphere has two equal curvatures, one round its equator, the other from pole to pole. Each single curvature generates the pressure t/r. Now consider the surface shown in Fig. 4. In principle this is still a cylinder, but it is curved, and no longer straight. The point X is on the inner face of the bend. Consequently, at this point there are two curvatures differing in sign. If the original circumferential curvature of the cylinder (in plumbing terms, the radius of the pipe) is reckoned positive, the backward bend of the surface at X must be counted as negative. But note that at Y *both* curvatures are positive: the surface at Y may not be a sphere but at least it is doubly convex to the exterior, as the surface at X is not. A surface with opposed curvatures may be called a 'saddle' surface (downwards on either side of the horse, upwards fore and aft of the rider). To see saddle surfaces produced by surface tension, make a loop of soft wire and dip it into soap solution to obtain a film across the loop. If the loop is flat, so that all parts of its perimeter will lie simultaneously upon a table, the film will also be flat, but if the loop is variously twisted, complex saddle surfaces can be obtained. As the pressure difference across the film in still air is zero we have everywhere $1/r + 1/r' = 0$, where r and r' are radii of curvature in two perpendicular planes, but r and r', if the twist of the loop is pronounced, will be seen to vary from place to place. Blowing on one side of the film will distort it so that there is no longer numerical equality of r and r' at each point.

Our argument to this point appears to raise the general expectation that cells which are free to do so will adopt spherical surfaces, or, where there is equality of pressure on both sides of a wall, flat interfaces. Where there is more serious interference between cells, preventing their walls from assuming spherical or flat figures, and imposing on them constraints equivalent to a twist in our experimental loop, interfaces between cells seem likely to have saddle curvatures.

Combining two cells

Two cells or bubbles constrained only by their contact with each other may present the appearance of Fig. 5, where there are three spherical surfaces such that $1/r_3 = 1/r_1 - 1/r_2$, as is necessary to give the partition-wall a curvature which will balance its tension with the pressure-difference across it. The first unequal mitosis in spores and zygotes of many plants will display a small cell bulging into a large one just in this way.

But have we completely solved the problem of combining two cells? Trials with compasses will show that we have not, and that our equation is not a workable basis for the actual construction of Fig. 5. Specifically, we have the radii of circles, but no locations for their centres. We have seen that the smaller cell must to some extent sink into the larger one, but we have as yet no way of fixing the point at which this movement is to be arrested.

Fig. 5 was in fact prepared by a geometrical trick, the mechanism of which is exposed in Fig. 6. Having prepared a line XY for the reception of our three centres, we choose an external point Z and draw ZP_1 not quite perpendicular to XY. Using Z as centre we draw a working-circle, two radii of which are then struck off as chords so that we may draw ZP_2 and ZP_3 at angles of 60° on either side of ZP_1. Drawing circles through Z centred on P_1, P_2 and P_3 gives us the construction required. This procedure is based on the consideration that the tensions in the three walls meeting at Z are to be equal, that consequently the angles at Z are to be 120°, and that three circles will meet at 120° when their radii meet at 60°.

The theoretical implications of this construction are of the utmost gravity, because even in the (biologically quite rudimentary) task of placing one cell in contact with another we shall find clear indications that the present train of thought is not one which we can reasonably expect to

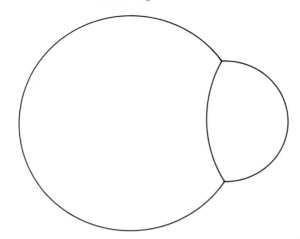

Fig. 5. Equilibrium of two unequal bubbles, a median section
through the assembly.

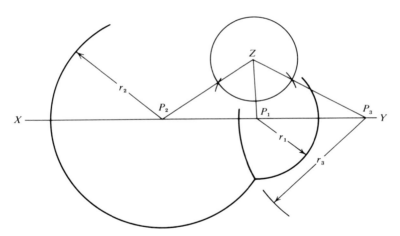

Fig. 6. Equilibrium of two unequal bubbles, geometrical construction
for finding centres and radii (see p. 18).

develop into any general theory of tissue construction. Consider firstly that
the procedure in Fig. 6 did not even begin with stated values of r_1 and r_2,
but with the arbitrary choice of the point Z. A natural response might be
to seek a solution to this specific mathematical difficulty and to deduce the
position of Z from given values of r_1 and r_2, but the biological relevance
of such an exercise is questionable in the extreme, because r_1 and r_2 are

not quantities to which obvious biological significance can be attached. They are not the radii which the cells possessed as separate units before they were joined, and they are not measures which bear any simple relation to the volumes of the cells.

These are serious weaknesses in our theoretical development, and they have become visible even before we have encountered the problem of saddle surfaces. The apparent simplicity of the two-cell problem, for what it is worth, rests entirely upon the fact that the point Z lies upon a ring-shaped seam in the outer surface of the assembly. This seam is a perfectly flat loop, and permits the interface to take a spherical form. That this state of affairs can survive the introduction of even one more unequal bubble into a two-bubble model is decidedly less than self-evident, while it is disagreeably obvious that the introduction of the fourth unit must automatically bring a higher order of mathematical difficulty. With three bubbles, all their centres must at least lie in a common plane, but with four this is no longer necessarily so. Furthermore the introduction of even the third bubble initiates a state of partial indeterminism, for three bubbles may be joined one-to-one in a row, or they may constitute a triangular cluster, every one in contact with both the others. Finally we have to note that every addition will change all the previous dimensions of the system. If in Fig. 5 we bring up a third bubble, then even if it touches only one of the original pair *all* the radii r_1, r_2 and r_3 will assume new values.

Upon any realistic assessment of all these matters we can only conclude that our solution of the two-cell problem is inherently incapable of being extended to cover in detail the problems of continuous tissues. We may usefully go a little further in considering the stability of limited assemblages of cells, but we must not allow this work to degenerate into a pursuit of the infinite by easy stages. The study of extended tissues has its own proper starting-point and is not to be approached by advancing from two cells to four, eight, and so on in ever-increasing powers of 2, or even by adding cells one at a time.

The biology of the two-cell contact

Our analysis of the two-cell problem shows that the smaller cell must be expected to have the higher internal pressure because of the sharper curvature of its surface, and that the action of surface forces will draw the smaller cell, to some extent, into a penetration of the larger one. It is at least a reasonable presumption that this situation, if not the primary cause, has been a contributory factor in the evolution of the many biological relationships occurring at the meeting of unequal cells (sexual fertilisation, parasitic invasion, ingestion). In continuation of the same line of thought we may also note, for example, that with equal wall tensions and equal

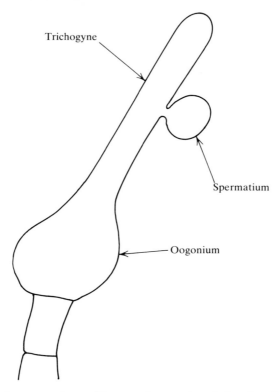

Fig. 7. Representative proportions of the sexual organs in red algae (applicable also to some fungi). The end cell of a filament enlarges into an oogonium with a narrow receptive extension or trichogyne. A male gamete or spermatium attaches itself to the side of the trichogyne and opens a pore through which the male nucleus passes. Although quite different conditions exist in some of the plants, 'normal' dimensions may be taken as about 5 μm for diameter of spermatium and trichogyne and about 10 μm for the wider part of the oogonium. As a guide to orders of magnitude, to give the spermatium an internal pressure of one atmosphere (and thus a pressure advantage over the female cell of half an atmosphere), the walls of the cells would have to be credited with about twice the tension which exists in a free water surface against air. This may be compared with an average car tyre, the tension of the tyre wall being about 200 000 times that of a water surface for an internal pressure not amounting to 2 atmospheres.

radii, the pressure in a spherical spermatium would be just double that in a cylindrical trichogyne (Fig. 7). Not, in all probability, a decisive comparison, but certainly one conducive to transference of material in one direction rather than the other.

It must not be supposed that the pressure relations which we have discovered are avoidable, or that it might be possible for a cell in some way to find protection from the power of surface tension. Most biologists, in their early studies of embryology and life-cycles, have made extensive use of simplified diagrams representing cellular constructions, and have thereby acquired a habit of drawing configurations which cannot survive critical examination. In the present context it may be useful to consider Fig. 8, a conventional representation of a pollen grain in which a small

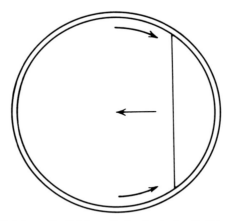

Fig. 8. Unequal division of a thick-walled spherical cell. The thin septum would be stable only if its edges were anchored immovably to the pre-existing rigid wall. If free to move, the new wall slides and bends as indicated by the arrows.

generative cell has been cut off. Because we may treat the spherical outer wall as rigid, there can never be exact agreement with the outline of Fig. 5, where the outer walls were treated as liquids. May not this rigidity of the outer shell give the two cells an opportunity to opt out of our expected pressure difference and maintain a flat interface? Pressure originally was the same at all points within the sphere, so why should it not remain so?

The answer lies at the edge of the small cell: we are trying to make this cell a plano-convex lens with a sharply bevelled edge meeting the rigid wall along a circular line. If the wall along this line could be developed, prior to any cytoplasmic separation, as a firm and immovable anchorage for the edges of the interface, then indeed the geometry of Fig. 8 might be preserved. But plasmolysis experiments and mitotic studies combine to show the implausibility of any such concept, and if the hypothetical sharp edge of the small cell is at any stage detachable from the outer wall it will be forced into retreat before a peripheral advance of the larger cell. As the small cell withdraws at its edges it must bulge at the centre, and the interface will become spherical in much the same way as in Fig. 5, though with some difference in detail. As a last resort, and in a final attempt to avoid the establishment of a difference in pressure, it might seem possible to argue that the interface could be unstressed, and not in tension at all. Mathematically this is a perfectly valid solution, but it is important to appreciate its consequences by reference to engineering principles of structural design. In the real world no component is ever the right length to fit a particular gap in a framework. Either it is too short, must be stretched to fit, and becomes a stay, or it is too long, must be squeezed into place, and will then serve as a strut if sufficiently rigid to do so. Flexible components like wires and aircraft skins are only kept in shape by their tensions, and a neutral condition of zero tension and zero deformation exists in them only as a transitory mathematical abstraction, a momentary prelude to visible structural collapse. In short, it is not realistic to attribute a complete loss of tension to thin and flexible membranes which are preserving smooth and unwrinkled contours.

Three cells

The triangular juxtaposition of three equal cells (Fig. 9) is a specially simple and symmetrical pattern, with three spherical outer faces and three flat interfaces between cells. The construction can cause no difficulty if the need for 120° intersection of walls is kept in mind.

At the centre of the figure it may appear that we have an entirely new feature in the meeting of three cells: this is represented in our plan as a point, but in the solid it must, from considerations of symmetry, be a straight line perpendicular to the page. It is easy to approximate this effect in a bubble model. We may call such a line an edge: it is the junction of three walls and in this particular instance it is straight because it is the junction of three planes. It is also the junction of three spaces.

But if we now recognise the external world simply as another space we may generalise the concept of an edge somewhat further. With two cells (Fig. 5) we had a single circular edge between the three spaces of cell, cell, and outside world. With our symmetrical trio (Fig. 9) we have, on the same basis, four edges. In reality the only new geometrical concept which appears with the addition of the third cell is the vertex, which appears as a junction of four edges, six walls, and four spaces. With three cells symmetrically disposed we have two vertices, whereas with two cells we had none.

We shall find these simple numerical relationships to be capable of indefinite extension to any number of cells. As long as surface tension governs the system it is impossible, for example, for four cells to meet at an edge or for five cells to come together at a vertex.

Besides using the three-cell model to introduce the concepts of edge and vertex, we may profitably pause to consider what will happen to it when the cells, instead of having equal volumes, are all of different sizes. Evidently the three cells must all have different internal pressures so that none of the interfaces can be flat. Let us guess, from our appreciation of the two-cell model, that the interfaces of the three are now to be parts of spheres. Obviously the central edge can no longer be straight. What kind of line will be formed by the central intersection of three spherically domed

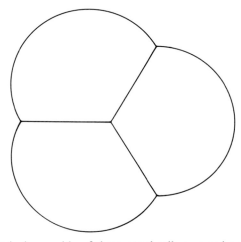

Fig. 9. A symmetrical assembly of three equal cells grouped around a central edge; the centres of cells are in the plane of the page.

interfaces? Simple consideration will show that no such line exists: the proposed intersection of spheres is a geometrical impossibility. For consider only two unequal spheres. Their only possible intersection is in a circle standing perpendicular to the line joining their centres. Consequently a third sphere to intersect jointly with our first pair must everywhere meet this circle, and it can only do this if its centre is on the line through the centres of the other two. With two cells (Fig. 5) we could make three unequal spheres meet in an edge because the three centres could be placed upon a common straight line. In the new problem we are unable to do this. Whatever its shape may be, the central edge between three unequal cells certainly is not an intersection of three spheres. And, by symmetry, unless all the interfaces are spheres none of them can be.

We have, in fact, made our first encounter with the more complex surfaces in which there is some element of saddle curvature, and from the biological point of view it does not appear that the exact mathematical delineation of curves and surfaces can be profitably carried further. The introduction of the third unequal cell marks for us the end of the road in this particular direction, because the mathematical refinements immediately overstep the bounds of practical observation and measurement. It would be pointless to calculate adjustments too small to be capable of verification from specimens.

Four cells

Juxtaposition of four cells is interesting to us as representing the level of complexity at which cell-centres cease to be necessarily coplanar. This we

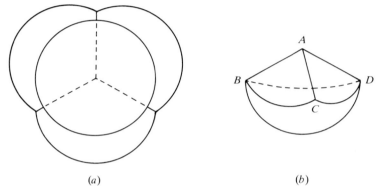

(a)   (b)

Fig. 10. (*a*) Four identical cells in the only arrangement which is possible without an intercellular space. Any three of the cells being placed upon a common level, the fourth cell will rest centrally over the top of them. (*b*) One cell extracted from the assembly. *A* is the vertex at the middle of the tetrad, and is the summit of a blunt triangular pyramid, the flanks of which (*ABC, ACD, ABD*) are flat because the pressure in every cell is the same. The inclination of these flat surfaces causes their curved edges (*BC, CD, BD*) to form bays or scallops, reducing the area of the convex exposed face of the cell.

have already foreseen, but the four-cell problem has also a more subtle significance as the starting-point for the complex interplay of shape and physiology which forms the main theme of this book.

If we accept the principles drawn from the three-cell model there can only be one perfectly symmetrical compact arrangement of four cells, because four spaces can only fit round a vertex. Round an edge there is room for three spaces and no more. Development of this idea gives the tetrahedral disposition of cells familiar to all biologists in spore-tetrads and the like (Fig. 10). With equal cells (and we have no need to consider any other state) each cell has four vertices, six edges, three plane faces and one spherical face. With this tetrahedral configuration we need have little further

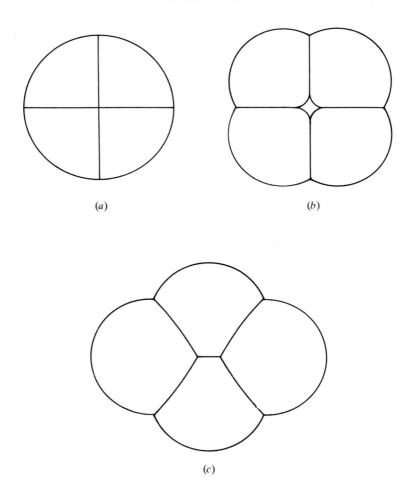

(a)   (b)

(c)

Fig. 11. Stability considerations relating to plane groups of four equal cells. (a) A conventional representation of a spherical cell twice divided. This configuration is so unstable as to be unobtainable in non-rigid systems. (b) A ring of four cells with a central space. This collapses instantly with bubbles, but is highly stable for cells if once established. (c) An ideally stabilised bubble model. Shown in section, and geometrically complex. None of the curved surfaces is a part of a sphere, and the volumes of the cells are not proportional to their areas in this diagram. Biological examples of this configuration are (always?) very imperfect for reasons discussed at p. 34.

concern, and any effort directed to the evaluation of angles etc. would be quite misplaced in a biological text.

Of incomparably greater significance is the disposition of four cells in a solid group with their centres in a plane or nearly so. Under the laws of surface tension and compression models, perfect symmetry is here impossible. For if we refer to Fig. 9 where three bubbles meet at a central edge we can see that there can be no way in which a fourth unit can ever intrude to the centre. Biologists are all familiar with diagrams like Fig. 11a, customarily introduced as the result of a spherical zygote having divided twice at right angles, much as the earth might be divided by two cuts, one on the meridians 0° and 180°, the second on the meridians 90° E and 90° W. Fig. 11a therefore implies the existence of a four-cell edge. If we accept this as a statement of fact, the simplicity of our theoretical development is at once destroyed, and we shall at the same time lose the close and helpful contact with inanimate models which has until now been supporting our argument. Juxtaposition of four units at an edge is unobtainable with bubbles, and in solid compression models it is obtainable only by careful and deliberately planned manipulation of the uncompressed spheres at the commencement of the experiment. The lesson of models is plain: the four-unit edge is not a natural phenomenon at all, but its emergence (or at least the emergence of an approximation to it) can (sometimes!) be forced upon the test materials by a sufficient exercise of human skill.

We have every reason to resist to the utmost any proposal that our scheme should be spoiled by the introduction of such a monstrosity as the four-unit edge. Nor, upon more critical examination of specimens, does there seem to be any necessity for such a drastic step. There are two ways of resolving the difficulty. In Fig. 11b we allow the four cells to separate slightly at the centre, so as to form a ring. This restores the simplicity of our theoretical development, and it is of course of no consequence to us how small the central opening may be, or by what type of extracellular material it may come to be occupied. In Fig. 11c we permit the objectionable four-unit edge to segregate into two acceptable three-unit edges a little distance apart. Careful examination of suitable biological specimens goes a long way to confirm, though it can never conclusively prove, that one or other of these interpretations can always be legitimately applied. So far as any general theory of tissue development is concerned, it is only the resolution adopted in Fig. 11c which need occupy us further.

What we have discovered is that four cells of equal volume, fitted into a continuous coplanar array, are unable to maintain a uniform morphological status: physical forces will compel the cells to differentiate, two and two, into structurally distinct categories. We have in Fig. 11c two 3-cornered cells or 3-gons, and two 4-gons. Furthermore the categories differ not only in shape but in environment; for example a 4-gon touches two 3-gons, whereas a 3-gon does not touch any other 3-gon at all. Many theoretical consequences will follow. If, for example, we suppose the four-celled assembly to grow by peripheral extension, then the 3-gons will be endowed with a higher relative growth rate, because these two cells are further from the centre, and have a larger share of the outer boundary, than their more centrally situated 4-gon neighbours. A hypothesis of uniform peripheral growth thus generates an expectation that the next mitosis may be somewhat advanced in the 3-gons and relatively regarded in the 4-gons.

Cells inside a rigid wall

We have so far provisionally assumed that all cell walls retain equal, and essentially fluid, physical properties until the whole arrangement of the system has been finally settled. The solution of the two-bubble problem depended on wall tensions remaining equal so that three walls should intersect at 120°. Biologically speaking, the exact opposite of this situation would be one in which cell walls achieved total rigidity within a single mitotic interphase. The problem at the next mitosis is then merely one of arranging a fluid partition across a chamber of specified but unchangeable shape. In the junction of a fluid wall with a totally rigid pre-existing one, it is easy to see that the new wall must come to rest so as to meet the old one perpendicularly, for if this condition is not satisfied (Fig. 12) the fluid anchorage will be subject to a lateral pull, and will slide along the old wall.

Using this principle of perpendicular attachment, it is easy, up to a point, to investigate theoretically the stable shapes for liquid partitions in rigid-walled cells of various forms such as cubes and tetrahedra, and even very complex problems of this kind can be approached by the use of models. We have therefore an extensive legacy of published discussions on the division of cells of particular shapes, supported by calculations and experiments, and periodically brought together into comprehensive reviews, of which the best is probably that by Schüepp (1966). Some of the examples studied have an immediate and obvious relevance to specialised biological situations. In relation, for example, to the apical cells of ferns and mosses it may be well worth while to consider in detail the stability conditions applicable to the development of a liquid septum in a rigid tetrahedral chamber.

However, in attempting to develop a general treatment of more ordinary tissues it does not appear that we can expect any major contribution to derive from this field of enquiry. By orderly presentation and by giving mathematically correct solutions for the partitioning of various shapes of chamber, it is possible to convey an impression of methodical approach to fundamental biological problems, but this impression, upon more critical examination, turns out to be almost totally spurious. For one thing the shapes customarily considered make up a highly artificial selection: the division of a geometrically perfect cube or spherical quadrant is not an operation from which one can see any obvious way forward to higher degrees of biological reality. Then again it is a manifest oversimplification to express a result in terms of one solution based upon minimal area of the newly formed wall. By calculation, or by use of a suitable model, an

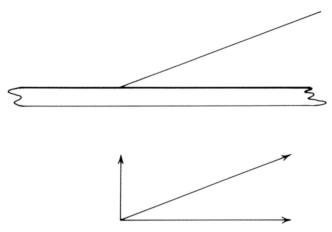

Fig. 12. Junction of a tensed fluid mitotic partition with an old rigid wall. Tension resolves into a component perpendicular to the old wall, which will have no visible effect, and a component parallel to the old wall, which will tend to cause sliding.

ideal mode of partitioning any given shape of chamber can no doubt be obtained, but the value of this is largely destroyed by the considerations introduced earlier (p. 14) about the polyhedra generated in compression models. Mitosis in an isotropic tissue must ordinarily be the bisection of a solid figure approximating to a sphere, and the expected partition is therefore an equatorial plane. For the actual partition in a given polyhedron there will be an infinity of different positions, of which only one can yield the absolute minimum area of new wall. But the range of variation in new wall area even for large changes in the position of the wall can only be rather trifling: there must be a multitude of differing orientations in which the new wall would contain more strain energy than it would have done in its 'best' position, but only by some quite negligible amount which it is not reasonable to suppose could materially influence the behaviour of a cell.

From this point of view, the best existing discussions of these matters are incomplete, and show a lack of understanding of the mechanical principles involved. It is not necessarily fruitful to distinguish a stability position in a system without at the same time evaluating the restoring force by which any departure from the ideal state is to be corrected. If the restoring force is very small, the equilibrium position becomes a theoretical abstraction, and will hardly ever be observed. As regards equal division of a given cellular shape, the restoring force which impels the nascent septum towards its ideal position of rest is limited by the departure of the

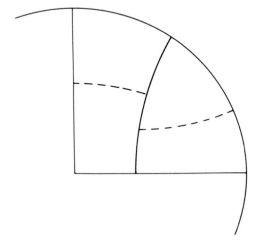

Fig. 13. Two successive stages of division, firstly by continuous lines, secondly by broken ones, in a circular quadrant, arranged on the basis that each partition is to have minimal surface and to solidify completely before the next division. A highly artificial scheme which may nevertheless have a certain significance for fern embryos etc.

parent cell from sphericity, and as we shall find reason (p. 64) to think that a dividing cell is likely to be a 17-hedron, it is only too probable that non-ideal septum orientations will pass uncorrected.

There are perhaps two results which are in some measure capable of being incorporated into a general theory of tissue structure. Firstly we may consider the division of a circular quadrant, that is to say of a cell shaped like a quarter of a Petri dish. On surface-tension principles the division is as shown in Fig. 13 and it is possible also to indicate the general character of the next divisions to be expected. Although this is a very artificial scheme, because we have every reason to reject the quadrant as a normal or even possible shape for a cell, it offers points of comparison with various embryos and discoid structures and carries forward the idea that increase in cell number must automatically lead to a process of morphological differentiation.

Of more fundamental interest is the introduction by Giesenhagen (1909) of a liquid model which is used dynamically rather than in a static condition. In a tall cylindrical jar, two immiscible liquids lie one over the other with a plane interface. When the jar is slowly tilted the interface becomes oblique to the axis of the vessel but remains flat. If now the jar is suddenly restored to the vertical, the interface is thrown into a sinuous

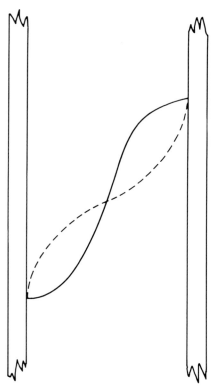

Fig. 14. Transient curvatures of liquid interface in Giesenhagen's model. In a cylindrical container the curvature is always compound: the edge of the interface has a different shape from a section across its centre.

curve (Fig. 14) reminiscent of various oblique septa in filamentous structures of plants, and notably in the rhizoids of mosses.

Rightly considered, Giesenhagen's model would seem to illustrate the true nature of the connection between cell shape on the one hand and cell-wall physics on the other. The essential new factor is the introduction of a scale of time. Suppose the inclined jar to represent the formation of a septum in an oblique position through some inclination of the mitotic spindle. Surface forces in the oblique wall, if given time, would turn the wall into a transverse position with consequent reduction of area and release of strain energy. This would imply perfect liquidity, and may be modelled by slow restoration of the jar to the vertical. But when the jar moves too fast, the contents cannot immediately follow. The properties of cell walls very largely conform to the requirements of such a scheme: under sudden stress a wall behaves as a solid, even perhaps to the point of rupture, yet the same wall, on the time-scale appropriate to tissue growth, may extend considerably without showing any resultant rise in tensile stress, which we earlier (p. 15) saw to be a characteristic of liquid films. The justification for any discussion of tissues in simple terms of surface tension is drawn in great part from the slowness of the phenomena involved. Giesenhagen's experiment is a reminder that irreversible solidification may set in before the principles of fluid equilibrium have fully asserted themselves. In the terms which have been used above, the septum in a moss rhizoid is oblique (and so not in its minimum-area position of equilibrium) because the restoring force was insufficient to correct, within the time which was available, the displacement initially imposed by the mitotic spindle.

Viscous resistance to cellular movement

When, in a surface-tension system, bubbles or cells move from an unstable configuration towards some more stable arrangement, there will be a viscous resistance and a viscous power loss. (There will be certain other power losses, which we can afford to disregard.) Energy used to drive a body through a liquid medium against the viscous drag is dissipated irrecoverably as heat. There is, consequently, no possibility that mathematically ideal minimum-surface conditions can ever be realised in material systems; a great deal that has been written about the 'theoretical' basis of cell-shape assumes that a tissue can operate as a perpetual-motion machine, miraculously exempted from the laws of thermodynamics. Such discussions have no claim on the attention of any serious student.

The most optimistic allowance which we can make for ease of movement in a liquid is an assumption of laminar flow with a Newtonian law of viscosity. This leads at once to a very striking scale effect, for which examples are readily available in naval architecture. Stop engines, and a big ship will run for miles, whereas a rowing-boat shows an obvious reduction of speed even between successive strokes of the oars. This is because the kinetic energy of a moving hull is proportional to displacement tonnage (and therefore to the cube of length) whereas the viscous drag is related to wetted skin area (and therefore to the square of length). Conditions are disproportionately unfavourable to the economical propulsion of smaller vessels. At cellular dimensions and speeds, the momentum of a moving body is no longer sufficient to achieve any measurable penetration of a medium as viscous as water. The cilia of a protozoon must beat out of phase because if they were synchronised every return-stroke would drive the animal astern. A unicell cannot 'free-wheel' or 'coast' between successive impulses as a human swimmer may do. The situation of the protozoon is more nearly comparable with that of a large animal burrowing through sand: continuous movement demands continuous propulsion.

Viscous power loss is proportional to the viscosity coefficient, and to the square of the speed. The power-source for the adjustment of any unstable arrangement of bubbles in a surface-tension model is the surface energy of 'vanishing' bubble surfaces. The available horse-power is therefore proportional to the rate of net diminution of surface area, so area must be differentiated with respect to time, and power-output is proportional to speed of collapse, *not* to the square of the speed. If, therefore, we set up identical unstable bubble systems in two liquids of similar surface tension but different viscosities we have the relationship:

(Speed of collapse) × (Viscosity coefficient) = Constant.

The reader will perceive that this is the expected result: it is simply the fundamental equation for a perfect viscometer.

For a biologist there can be few forms of literature more thought-provoking than a table of viscosity coefficients: it is really astonishing how the values go up as liquids become, by intuitive assessment, 'thicker'. We shall not be acting at all unreasonably if we speculate that quite young cellular interfaces may have viscosities exceeding 100 000 times that of water. With speeds of movement reduced in proportion, it is perfectly childish to argue that cells 'must' conform to idealised minimum-surface geometry. The question which is really at issue is whether a cellular configuration will move into recognisable conformity with the surface-tension theory before it is swept out of existence altogether, either by mortality or by a new outburst of mitotic activity.

The biological relevance of bubble models therefore depends on degrees of instability. If we are able to say, of some bubble configuration, not merely 'This will be unstable' but 'This will have a specially high collapse-rate', then, and then only, will it be reasonable to expect general biological exemplification of the corrective movement. How can we decide which geometrical changes will be quick and which will be relatively slow? In reality it is easy to attain a rough appreciation of this matter, quite sufficient for ordinary biological purposes.

Consider a ring of four equal bubbles around a central space (Fig. 11*b*). This is unstable, and four bubbles placed in this manner will fall into the middle and obliterate the opening, but the collapse-rate is quite low. Let each bubble move a little towards the centre, and there will be a contraction of its wall round the circumference of the ring, and the drive for the movement will have to come from the tension-energy which was previously stored in the disappearing part of the surface. But as a cell moves inwards it is compressed between two converging radii. There must consequently be an increase in the girth of the cell between the outer edge of the ring and the central opening, and its interfaces with its two neighbours must also be enlarged. Actual annihilation of surface is therefore a relatively small item in the account. The energy which is liberated, and made available to perform external work, is merely a residue, a credit balance arising from large internal transactions, all of which are subject to heavy taxation by viscous power loss. A ring of four bubbles will collapse as a pure surface-tension theory would dictate, but a ring of four cells will show very little inclination to do the same: its position is analogous to many engineering examples in which friction outweighs the forces which can be brought against it. In a frictionless world every bolt would spin right out of its nut at the slightest pull: careful observation will reveal that it is actually not customary for nuts and bolts to behave in this way, and the idea of non-viscous cells is no more realistic than the concept of frictionless screw threads.

The prospects of corrective adjustment in any particular case may be notionally evaluated by the fraction:

$$\frac{\text{Rate of net area reduction}}{\text{Rate of internal redistribution of area}}$$

and when this quantity is small there can be no realistic expectation that idealised minimum-surface geometry will ever be attained.

Without formal mathematical analysis it is quite easy to see the general application of this principle to various geometrical figures. Consider, for example, cells forcibly placed in the situation of Fig. 11a. Surface forces will prescribe a movement towards the pattern of Fig. 11c. During the early stages of this movement the instability fraction is very large, indeed it is initially, for an infinitesimal period of time, infinite in the strictest mathematical sense. The very first separation of a four-cell edge into two three-cell edges is a pure energy release, unencumbered by any need to provide power for concomitant rearrangements in remote parts of the system. But before the arrangement of Fig. 11c can be reached, the instability will have been reduced almost to zero. It will be seen, for example, that Fig. 11c shows considerable elongation of the four-cell group. This elongation can be attained only by huge internal exchanges of surface energy, all of which must be driven by a power-source which may fairly be described as negligible. In biological conditions, therefore, the movement from Fig. 11a to Fig. 11c will always begin but is never likely to be completed.

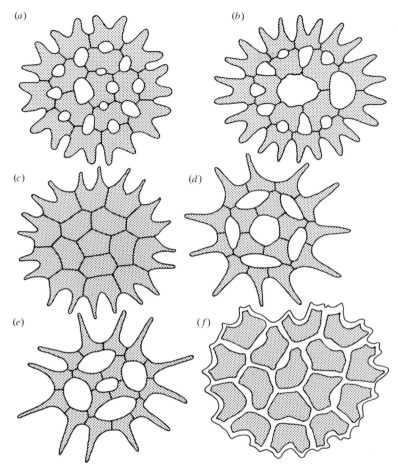

Fig. 15. Selected sixteen-cell colonies of *Pediastrum*. (*a*) A potentially symmetrical arrangement of *P. asperum* spoilt by failure to suppress innate shape of central cell. (*b*) *P. clathratum* with a central space is unsymmetrical, owing to arithmetical incompatibility between rings of 5 and 11. (*c*) *P. boryanum* achieves symmetry by full compression with a central cell, but note that this design is incapable of being enlarged to fit a 32-celled colony. (*d*) *P. triangulum* in this specimen fails to find a symmetrical design. (*e*) Another colony of the same species manages, by virtue of radial elongation of its cells, to join two rings of 4 and 12. (*f*) *P. angulosum* disguises imperfections of fit by producing interstitial material, but the increased bulk makes it no longer possible for five cells completely to surround the central one, and a peripheral cell falls into the gap.

Concentric plane arrays of cells

The search for informative models of tissue structure leads inevitably to various exercises in which identical units are spaced out upon a surface by some kind of mutual interaction, with or without some further external constraint. Many expedients are available: for example magnetised needles may be struck in cork discs, all with similar poles uppermost, floated on a bath, and allowed to take up positions around a central attracting magnet under their own mutual repulsions. Ingold (1973) has used floating wooden discs in a large funnel from which the water can be progressively drained so that the sloping walls force the units together. Whatever the mechanisms employed all such experiments are essentially compression models in two dimensions rather than three.

The coenobial alga *Pediastrum* is of interest in this connection (Harper, 1917, 1918). Cells are formed in some number which is usually a power of 2, are initially free-moving, but are then required to dispose themselves as nearly as possible in a circular plate (Fig. 15*a–f*). Comparison of the various species, and of colonies differing in number of cells and degree of compression, though giving a useful insight into the mechanics of compression, clearly cannot lead to an adequate understanding of tissues with a more normal mitotic regime.

Extended plane compression models

In various ways it is easy to find, or, if desired, to make, plane compression models not restricted to a few units but extending indefinitely in such a way that any area which is selected for detailed observation can be treated in practice as part of an infinite array. To take but a single example, Lewis (1931) made use of one of the plates prepared commercially for a long-extinct process of additive colour photography. The network here consisted of resin-globules, dyed in contrasting colours, and compressed into a single layer in the manufacture. Although such models can be used for illustrative purposes there seems to be no place for them in serious scientific discussion. Two-dimensional problems can be treated mathematically in such an exact and effective manner as virtually to eliminate the need for modelling experiments. In this respect there is a sharp theoretical distinction between work in two dimensions, for which the use of models offers no particular advantage, and the much more difficult field of three-dimensional study, in which it must be expected that compression modelling will form a permanent and indispensable division of the science.

3

The division of a surface

A tissue presents to the eye a pattern of linked polygons, so we need to know the mathematical laws which control the division of a surface into various connected n-gons, of which two are to meet at each wall and three at each corner. We can afford to permit unrestricted curvature of the sides, so that a 2-gon will be a possible figure. Any results we obtain will be equally applicable to surface views and to histological sections.

Initially we may avoid any consideration of the problems associated with the edge or boundary of our system. We shall simply suppose the tissue to extend beyond the range of our vision to an unlimited distance in all directions. The enquiry thus assumes a statistical character and the portion of tissue under our observation has to be accorded the status of a sample only. Our interest is shifting from the affairs of individual cells to the average conditions of a whole population. Such conditions, in respect of infinite populations, are theoretical abstractions. They cannot be observed, but only estimated, and we must keep in mind the statistical principle that estimates must be expected to display chance deviations in inverse ratio to the square root of sample size.

If we draw a network of polygons on a flexible and extensible surface, say a rubber sheet, stretching and bending will alter the distance between marked points in the diagram, but will not change the value of n for a given n-gon. For our present purpose, which is merely a study of the variation of n, we have therefore no need to stipulate that our surface should be flat. We shall be wise, however, for reasons which will appear later, to regard it as an open surface, that is one which does not join up in such a way as to enclose a central cavity. The construction of polygons on a closed surface, such as that of a ball, raises additional problems.

Proof that the average value of n must be 6

We begin by establishing a law of conservation, in that re-arrangement of cells neither creates sides nor destroys them, but merely transfers them from cell to cell. Suppose cells A and B (Fig. 16a, b) to move towards each other with the prospect of forming a common interface and so gaining one side each. If we are to permit any possibility of re-arrangement at all we shall

37

have to suppose that the system may pass instantaneously through the unstable position of a four-cell corner and so arrive at the condition of Fig. 16*b*. But now the anticipated gain of two sides by *A* and *B* has been balanced by a loss of two sides by *C* and *D*. The average value of *n* for the tissue is therefore unchanged. Any more complicated scheme of cellular movement will merely resolve itself into a number of repetitions of this unitary four-cell transaction. The required principle of conservation follows inescapably.

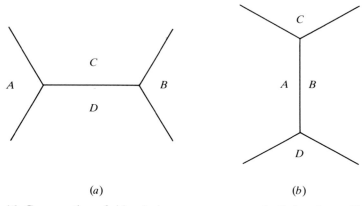

(*a*)   (*b*)

Fig. 16. Conservation of sides during rearrangement of cells in a layer. If cells *A* and *B* move towards each other they can gain one side each by establishing a new mutual interface, but only at the expense of cells *C* and *D*, each of which loses a side in the same transaction.

We now consider the addition of a new cell to the mass. The form of addition which is mathematically simplest is the expansion of an existing corner to accommodate a new 3-gon. The new cell has three sides of its own, and each of its neighbours registers a gain of one side. The total addition to the account for the whole tissue is therefore a package of (1 cell + 6 sides). If we do not desire a 3-gon, any other configuration which is geometrically possible can be secured by appropriate transference of sides under the rule of conservation. Reversing the argument to cover deletion or obliteration of a cell, we have the law that addition or subtraction of one cell carries a concomitant addition or subtraction of six sides, and this is sufficient to establish the general average of six sides per cell.

The exchange of sides resulting from mitosis

The division of a cell by a partition will most often take the form shown in Fig. 17*a*, where the partition is anchored on two different sides of the mother-cell. For completeness we may also include the rarer condition shown in Fig. 17*b*.

In either case the division increases the number of cells by one, so we know from the previous section that it must increase the total number of sides in the section by six. Examination of the diagrams will reveal the following generalisations:

(*a*) Mitosis in a mother-cell which is an *m*-gon results in a *total* endowment for the daughter-cells of (*m*+4) sides.

(*b*) Every mitotic division imparts two additional sides to cells not involved in the mitosis.

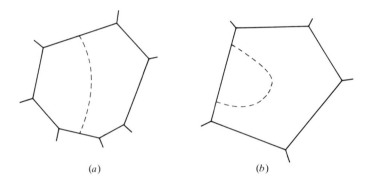

(*a*)   (*b*)

Fig. 17. Anchorages of a mitotic partition, (*a*) on different sides or (*b*) less commonly on the same side, of the mother-cell. See text p. 38.

The second of these conclusions implies that any tissue in which mitosis is to go on continuously must have a circulatory pattern of geometric change. An excess side (that is, a side additional to the number required to give any sample of cells its due average of six sides per cell) is an indestructible mathematical entity. If every mitosis liberates two excess sides into the non-dividing or interphase cells, the mean polygonal grade of the interphase cells will continuously increase unless there is some way of recycling the excess sides into future mitoses. A stable condition will only be possible if there is a dynamic equilibrium between the outward transference of sides after mitosis and the inward transference before mitosis. To find a precise mathematical statement of this equilibrium must be our next concern.

The steady state of a tissue

Let us assume total statistical uniformity in our tissue. By this we mean that observational estimates of any property of the tissue are to be independent of the time at which they are taken, independent of the place at which they are taken, and independent of the direction of the observer's course as he moves through the tissue to collect his sample. This condition may be defined as the *steady state*.

The most obvious property of a steady-state tissue is the existence within it of a standard average life-course for a cell from one mitosis to the next. Individual cells may depart from this standard, but they can only do so on a random basis. Similarly the distribution and orientation of mitotic figures must also be randomised. It may be helpful to express this a little differently, in terms of the difference which may exist between any two cells at a given time. We may choose two cells as strongly contrasted as the tissue will afford in such matters as shape, size, mitotic phase, and direction of the next impending mitosis. The existence of a steady state requires that the condition represented by each cell should be freely attainable by the progeny of the other. The concept of steady state will not encompass a situation in which 4-gons and 8-gons are allowed to propagate differentially, each in its own kind. Both figures are to be developed, if at all, only as passing incidents in *one and the same* pattern of geometrical circulation.

The steady state is therefore a very precise statistical abstraction and it will of course be a matter of observation to find how closely any real tissue may conform to it. The steady state is introduced merely as the simplest condition which could possibly exist: if we cannot understand the steady state we have little prospect of understanding anything more complicated.

Equilibrium of the steady state

We know from previous work that the general average number of sides per cell is exactly six, and that every form of action which is open to a cell in mitotic interphase is powerless to change that average. The possible transactions among interphase cells comprise:

(*a*) Exchanges, as shown in Fig. 16.

(*b*) Death or withdrawal of a cell with occlusion of the resulting space by neighbours.

(*c*) Intrusion of a cell from another plane of section.

In relation to the six-side average, each of these events is strictly neutral. It is therefore necessary to arrange for similar neutrality in the remaining portion of the cellular life-cycle, that is to say in the episode of active mitosis. The general average number of sides for all those cells entering or emerging from the mitotic transformation must be six. If m is the average for all mother-cells we have:

$$\tfrac{1}{3}\{m+(m+4)\} = 6$$

and $m = 7$. This is a necessary condition for the existence of a steady state. It is not however a *sufficient* condition (for one might have a tissue in which mitosis, initially the exclusive prerogative of 7-gons, was allocated progressively to 6-gons and 8-gons in equal numbers as time went on).

In average terms only, therefore, we find that a cell in steady-state tissue is born as a $5\tfrac{1}{2}$-gon, and divides as a 7-gon, acquiring $1\tfrac{1}{2}$ sides during the interphase owing to the mitotic activity of neighbours which select it as an anchorage for their partitions. The occurrence of fractions ensures that no cell can ever individually conform to this scheme: a $5\tfrac{1}{2}$-gon is not a possible figure and the most nearly equal division of a 7-gon will yield (5-gon + 6-gon). The observations of Lewis (1943*b*) will sufficiently exemplify the principle, but in fact there has been very little incentive for biologists to study cells in this way at the rather elementary level which our arguments have so far attained. The main comparison of theory with observation will come later (Chapter 4) when we have learnt to deal with cells as solid bodies and develop an equation for use in three dimensions instead of two.

It may cause concern that a pair of daughter-cells, in order to progress to 7-gon status and themselves qualify as mothers, must acquire three additional sides, although at their birth the number of sides imparted to the interphase tissue was only two. Is this not an imbalance which threatens the stability of the system? One mitotic cycle put two sides into the interphase cells, but three sides must come back if the next mitotic cycle is to follow the same geometrical pattern. Where can the extra side come from?

The key to the paradox lies in the timing of mitoses. If the two daughters of one mitosis were themselves to divide simultaneously, this would be a breach of the steady state, because it would establish a local peculiarity of behaviour, and also a fluctuation with time. Once we tolerate fluctuations and local peculiarities we have morphological differentiation, not a steady state. The three sides we have just mentioned will not therefore all be needed at the same time. Two sides would satisfy the needs of one daughter: after it has divided it will be able to lend a side to its sister to facilitate mitosis of that cell a little later. To understand the true balance we need consider only the sides by which a cell is in excess or deficit of the general average of six. A 7-gon has an excess of one, which can only have been borrowed from the rest of the tissue: somewhere another cell must show a corresponding deficit of one, directly attributable to the loan. When the 7-gon divides, two sides go into the surrounding tissue: that is to say, the original loan is repaid, and an equal loan is offered in the reverse direction. The history is:

(1 side borrowed) → (1 mitosis) → (1 side lent).

This balanced simplicity exists only for the 7-gon mother. For division of a 6-gon we have:

(nothing borrowed) → (1 mitosis) → (2 sides lent)

and for an 8-gon

(2 sides borrowed) → (1 mitosis) → (nothing lent).

These are individually unbalanced, but produce a balanced account when added together. From this point of view it is easy to see that any departure from the seven-side maternal average involves an uncompensated gain or loss of sides for the interphase component of the tissue.

The diversity of cells

Nothing in our theoretical development enables us to set limits to the individual variability of cells in a steady-state tissue. The scheme of accounting which we have just introduced would be satisfied if, for example, mitosis occurred in 4-gons and 10-gons exclusively and in equal numbers, but never in any grade of cell between. One has only to state a rather extreme hypothetical example to show that our mathematical treatment at this stage remains very incomplete: our geometry is far too tolerant of bizarre situations which are unlikely to be permitted in the biology of an organism. Simply as a matter of common experience, homogeneous and undifferentiated tissues do not give the impression that cells are very free to take up the most aberrant polygonal forms. Generally the 6-gon is not only the average condition but also the commonest condition. 5-gons and 7-gons occur also with substantial frequencies, but beyond these the rate at which higher or lower polygons can be found falls away very steeply. What are the factors which underlie this apparent restraint on the geometrically possible range of diversity?

Undoubtedly the dominant consideration is the relative uniformity of cell size which is implicit in the steady state. The mitotic cycle involves in principle an oscillation of cell volume with maximum and minimum in a ratio of 2:1. If our cells are in a sheet of constant thickness, the corresponding ratio in the linear dimensions across the sheet is less than 3:2. If our polygonal outlines are sections of solid cells approximating to spheres, the ratio of maximum and minimum diameter directly attributable to the mitotic cycle is little more than 5:4.

In a sheet of cells of uniform size, fluidity of the walls will, if sufficiently powerful, impose the total geometrical uniformity of a net of regular hexagons. This can be modelled by the use of large equal bubbles between parallel glass plates. Where the tensions in the walls are able to enforce to any extent a rule of 120° intersection at each corner, a high polygon such as a 10-gon can exist only by being much larger than its neighbours. The size-range attributable to the mitotic cycle is simply insufficient to support any wide diversification of polygons. Variation of a purely accidental kind (the exceptionally large mother-cell, the unusually small daughter) must be expected to increase the range a little, but even if we double the volumetric range to 4:1, the corresponding ratio of linear dimensions in a solid tissue is still only a little over 3:2. Furthermore, it has to be observed from the purely geometrical point of view that if the division of a cell is to be as nearly as possible an equal division, mitosis will act as a restraining influence on the diversification of forms. If, for example, division takes place in 6-gons, 7-gons, and 8-gons (i.e. three polygonal grades), the daughter-cells need not belong to more than two grades (5-gons and 6-gons).

These considerations lead to the expectation, imprecise in its quantitative expression but apparently quite secure as a general principle, that the more aberrant polygonal grades such as 4-gons and 9-gons can only reach very low frequencies of occurrence in any tissue which even approximates to the steady state.

Departure from the steady state

From the position which we have now reached, it is a simple matter to appreciate, if only in descriptive terms, the consequences which must ensue when a tissue departs from the steady state in which the average grade of mother-cells is that of a 7-gon.

Consider first the situation in which, on average, $m < 7$. We now have an excessive outflow of sides into the interphase cells, which will as a result be prematurely upgraded and hastened towards their own next mitosis. Mitosis thus becomes a contagious phenomenon, spreading through the tissue as a wave or pulse. It is easy to see that if we insist on a general average of $m = 6$, then the velocity of propagation becomes infinite: this is a prescription for absolutely simultaneous mitosis throughout the system. Construct a net of 6-gons and divide any one 6-gon evenly into two 5-gons. Two adjoining cells are now 7-gons. Under the rule of $m = 6$ these two cells are now post-mature and ought to be divided immediately, but if we are to do this and maintain the average of $m = 6$ our two 5-gons must also be divided and when these operations are complete we shall have several newly upgraded 7-gons to attend to, and so on without limit.

The only way of avoiding this result is to permit cellular dimorphism. If cells in polygonal grades lower than the 7-gon are to become disproportionately active in mitosis, this will exaggerate the tendency, already inherent in the steady state, for cells which are mitotically sluggish to become passive receptors of sides and to be upgraded in the scale of polygons. If cells progressively drop out of the mitotic cycle and submit to be upgraded into high n-gons, presumably with appropriate increase in size, then the diminishing fraction of the population which remains mitotically active will be able to maintain a lowered average value of m. Effectively the high n-gons have become a sink into which unwanted sides can be discarded by the active component of the tissue. It has to be noted that there is a natural limit to this process. If 12-gons and the like become so numerous or so large that they begin to have interfaces with each other, rather than with cells of much lower grade, new complications will present themselves.

Turning now to the possibility of a regime in which mitosis becomes disproportionately a privilege of high-n-gons, it is readily apparent that such a state of affairs can only be sustained for any length of time if the division of a cell is made unsymmetrical. For consider the situation of, say, a 10-gon. The rule of six-side average is sufficient to ensure that a 10-gon will be surrounded by other cells generally of much lower grade then itself. That even one immediate neighbour should be as much as a 9-gon is possible but not very likely. Equal division of a 10-gon produces two 7-gons, and unless the partition had an anchorage-point on a neighbour which was at least a 9-gon (which implies a conjunction of two rather improbable events), our original 10-gon has been extinguished without replacement. If mitosis is exclusively reserved for 10-gons and above, the degree of geometrical downgrading which is inherent in the equal division of high polygons will ensure the cessation of mitotic activity almost everywhere in no more than two or three cycles. But any high polygon is self-replacing if it divides so unevenly as to make one daughter a 4-gon. The prospect before us is therefore that of a select minority of high n-gons establishing themselves as permanent local features of the tissue, dividing unsymmetrically so as to surround themselves with an ever-increasing mass of low n-gons, few if any of which will every reach the geometrical status which we are now supposing to have become necessary for division.

This exhausts the possibilities which can arise from any general or over-all departure from the steady-state cycle of geometric change. The forms of differentiation which we have recognised are:

(*a*) Synchronisation of mitoses.

(*b*) Cellular dimorphism: mitotically active low polygons contrasted with inactive high polygons.

(*c*) Localisation of mitosis: active division confined to a minority of cells which can maintain their geometrical rank only by dividing unequally.

Most biologists will have little hesitation in discarding (*c*) as being simply inapplicable to any tissue within the ordinary range of experience. Except in early embryonic states and certain other rather specialised cases, extensive synchronisation of mitotic phases in different cells is evidently not a phenomenon to be invoked in a general theory of histological differentiation: systematic fluctuation of mitotic rate no doubt is widespread, but it produces little geometrical effect because it does not ordinarily attain such precision as would be required to coordinate the behaviour of a particular cell with that of its immediate neighbours. We must therefore also very largely set aside our theoretical case (*a*) above. Practically speaking, nearly all the geometrical problems which arise in the differentiation of tissues will be found to be comprised in the single remaining case (*b*).

Relationship of steady-state theory to observation

We are now in a position to plan the application of our theoretical concepts to the practical evaluation of tissue samples. Suppose that we have a mature tissue in which the division and enlargement of cells have been for some time discontinued. The very simplest situation which can be imagined (given that our tissue developed mitotically, which involves an absolute *minimum* size-variation of 2:1), is an instantaneous arrest of all further change in a steady-state tissue. We shall then see a tissue in which *all* systematic differences between various categories of cell are attributable to the mitotic cycle. We shall therefore have a predetermined relationship between the size of a cell and its geometrical status, because the polygonal grade 7 must refer to a cell twice as big as the grade $5\frac{1}{2}$, and so on. This line of argument could fairly easily be pursued to the point of direct comparison with actual measurements.

In fact it would not be worth while to seek verification at this stage, because we shall soon have more powerful methods at our disposal. In Chapter 4 we shall be able to test the steady-state theory by treating cells as solid three-dimensional bodies. Anticipating the results to be obtained there, we can say with confidence, as an observational fact, that many living tissues of the simpler kind are remarkably similar in their construction to the mathematically idealised steady-state tissue arbitrarily frozen at a particular instant.

In so far as there has been any significant departure from steady-state geometry, the arguments of the last section will lead us into a search for distinct classes or categories of cell, with a clear understanding that each of the categories we are able to distinguish may have a characteristic net excess or deficit of sides per cell, but only within the framework of a strictly balanced account for the tissue as a whole. Methods for this kind of enquiry will be developed in Chapter 7.

Polygons on a closed surface

A network of linked 6-gons can be indefinitely extended across a flat surface without any difficulty of construction: so long as the surface itself continues in a blank state before us we shall always be able to add another 6-gon to the edge of our diagram. If we perform the same construction on

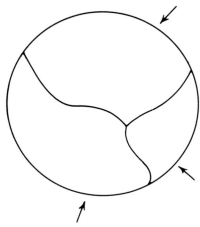

Fig. 18. A circle divided into three 3-gons can be plastically deformed into a hemisphere: if a similar hemisphere is added with its partitions alternating (arrows) we have six 4-gons covering a sphere.

a ball or an egg, a time must come when our advancing repetitive sequence of 6-gons no longer has much blank surface in front of it, but has to make an approach to some older part of the drawing. Shall we be able to make the join with 6-gons only? If we cannot achieve uniform coverage with 6-gons could we do so by using some other grade of *n*-gon?

To obtain a method for dealing with problems of this type, consider (Fig. 18) a circle composed of three 3-gons. Imagine this diagram to be concave, and join its perimeter to that of another identical figure, remembering to alternate the two rings of 3-gons so as to preserve the rule of three-cell corners. This converts every 3-gon into a 4-gon, so six 4-gons will cover a globe. From the principles established earlier in this chapter, any further elaboration of our globular network can only be the equivalent of introducing some whole number of 6-gons.

Therefore if we cover a closed surface with assorted *n*-gons to the total number of N polygons, making a solid figure which we may call an N-hedron, then the grand total of sides for all the n-gons must be $(6N - 12)$. The division of a closed surface is thus always subject to a fixed total deficit of twelve sides, which the conservation principle will allow to be very variously disposed. If we are resolved to cover a ball by using 6-gons as far as we can, then the join of the pattern can be effected by using twelve 5-gons, or six 4-gons, or by various mixtures, but never with 6-gons alone.

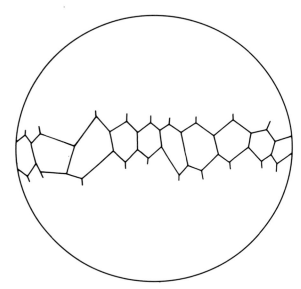

Fig. 19. A globe girdled by a belt of 6-gons, each of which exposes two free sides towards each pole. By symmetry, a net of polygons to fill either cap must have a deficit of six sides (see text p. 46).

If we wish to introduce 9-gons etc. this can be done, but there will then have to be appropriate overcompensation elsewhere to secure the necessary global deficit.

For any tissue displayed upon a closed surface the average number of sides per cell is therefore indeterminate. It is always less than six, but how much less depends on the number of cells. The rule is nevertheless a very simple one and for structures of roughly globular form nothing further is needed. Immediate biological applications can be found among the cells on the outer surface of an embryo, in the linings of cavities, in the sculpture or exoskeletons of various unicells, plant spores etc.

In dealing with more complex shapes it becomes expedient to divide the surface and to consider separately the deficits attributable to the various parts. To understand how this can be done imagine a globe which has been completely segregated into two polar caps by means of an equatorial belt of 6-gons, these being so disposed (Fig. 19) that each constituent 6-gon exposes two free sides towards each hemisphere. This arrangement is sufficiently symmetrical to ensure that each polar cap will have a deficit of six sides exactly.

Evidently we have here the foundations of an observational method: if we can find in our specimen any suitable ring of 6-gons, it appears that

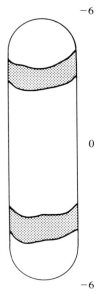

Fig. 20. A rod or tube, each of the shaded bands containing a symmetrical ring of 6-gons similar to that of Fig. 19. Distribution of the deficit is then as shown.

this will act as a barrier. We can be quite certain that a deficit of six sides is somewhere inside our cordon, and if we do not need to locate the deficit more precisely there will be no need to look at any of the enclosed cells individually. The existence of a suitable chain of 6-gons in a convenient place is purely a matter of luck, but this is only a minor technical difficulty which will be disposed of in the next section. In reality it is a simple matter to run a perimeter round any chosen region of tissue and determine the total net deficit or excess for all cells within that perimeter. We do not need to examine the entrapped cells individually, we do not even need to know how many of them there are.

We may now extend our treatment to more elaborate surfaces, using diagrams in which a shaded band represents a test perimeter equivalent to a symmetrical barrier of 6-gons. In Fig. 20 our globular figure has elongated into a rod or capped tube. Assuming for the moment that we can find an appropriate perimeter reasonably close to each end we shall be able to attribute a deficit of six sides to each end and a deficit of zero to the middle tubular section. The interesting new feature here is the emergence of zero as being apparently the deficit attributable to an

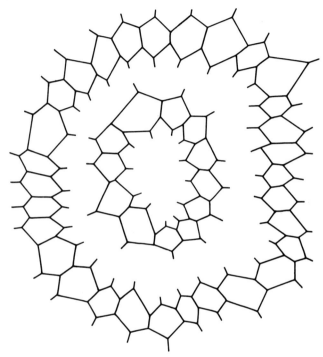

Fig. 21. Two symmetrical belts of linked 6-gons. A polygonal net to fill the interspace between two such rings will have zero deficit, as in the middle portion of Fig. 20. Infilling a single belt to the centre creates a deficit of six sides.

unbranched tube (ends not counting). In fact this is a geometrically correct result: between any two symmetrical rings of 6-gons, any infilling which can be constructed always averages six sides per cell. To test, prepare a diagram in some such form as Fig. 21, take a different colour of pencil, draw assorted *n*-gons to join the outer and inner rings, respecting the rule of three-cell corners, count sides, and average. By this result, the tissue on the surface of a ring or torus would also have zero deficit, for a torus is an unending tube (alternatively, refer to Fig. 21 and compare with a motor tyre: place the inner ring in the rim of the wheel, the outer in the tyre tread, and provide infilling for both side walls).

In Fig. 22*a* we proceed to a branched tube, previous work enabling us to attach values to the three ends and the straight tubular portions. As our whole specimen is still (by plastic deformation) interchangeable with a sphere, the total deficit is only twelve sides, so we must attribute to the

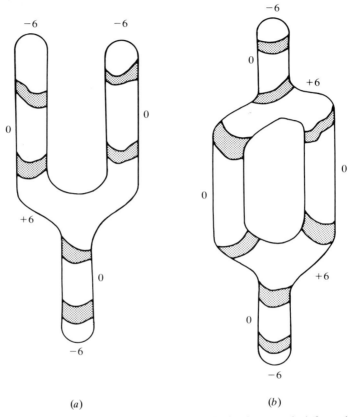

(*a*)   (*b*)

Fig. 22. Complex closed surfaces. (*a*) is a branched tube or rod, deformable into a sphere, and consequently with a total deficit of twelve sides, the distribution of which is given on the assumption that suitable barriers of 6-gons exist in the shaded areas. (*b*) is not deformable into a sphere but it *will* deform into a torus, for which zero is the correct total deficit. It is left as an exercise for the reader to design a surface with an overall *excess* of sides.

point of branching an *excess* of six sides. By symmetry, the anastomosis of tubes (Fig. 22*b*) must be similarly treated.

We thus obtain a simple set of rules applicable (Lewis, 1933*a*) to such matters as the lining of capillary blood-vessels. A little less obviously, similar reasoning can be brought to bear upon prominent relief features on an open surface. Suppose a surface to be developed into numerous raised multicellular tubercles or papillae, or to be diversified in the reciprocal manner with deep tubular pits. The top of a papilla or the bottom of a pit is then geometrically comparable with the end cap of a tube or rod and may provisionally be assigned a deficit of six sides. The total deficit of an open surface being zero, a counterbalancing excess of six sides must now be attributed to the general surface surrounding each of the relief features. It is entirely logical and consistent that this should be so, for the part concerned is exactly comparable with the site of branching in a tube (think of the region between two fingers of a glove, turning inside-out as required).

Perimeter check

Evidently there are many biological contexts in which we may wish to ascertain what excess or deficit of sides may exist in some given region of tissue. Having regard to the peculiar rules of this branch of geometry, we shall often have little interest in the number of cells involved, and even less interest in the shapes of these cells considered individually. The notion of locating an excess within a test perimeter is therefore attractive, but we have to emancipate ourselves from the need to find a chain of 6-gons on every occasion.

Suppose for the moment we happen to have a symmetrical ring of 6-gons available, and consider the outer perimeter of that ring. This perimeter is a line from which cell walls branch off alternately to the outside and the inside. Let E be the total count of external walls meeting the perimeter and I the count of internal walls. At present $I = E$, and we know from previous work that the total enclosed deficit is exactly six sides. Imagine a mitosis contiguous to the perimeter, and adopting the perimeter as an anchorage for the new wall. The mitotic partition then constitutes an addition of 1 to either I or E, while a cell on the other side of the perimeter gains one side owing to the division of one of the existing sides by the anchorage-point of the partition. So as E increases, sides are injected into the enclosed tissue, as I increases, sides escape from the enclosure into the external tissue. We have:

$$\text{Total enclosed deficit} = I - E + 6.$$

This result is perfectly general, and more flexible in biological application than the treatment by Graustein (1931) to which a number of biological writers have referred.

Routine laboratory use of this principle will not often go beyond the elementary case of a single loop of test perimeter enclosing a small patch of tissue. More complex operations (e.g. girdling a tube in two places to check the condition of a cellular lining) require great care in defining what is meant by 'internal' and 'external', but will otherwise present no theoretical difficulty if they are ever needed.

The polyhedral cell

In the three-dimensional analysis of tissue structure, we have to deal concurrently with two levels of statistical sampling. The tissue consists of various assorted N-hedra, and the faces of any single N-hedron, or for that matter of any specified assemblage of them, will comprise similarly assorted n-gons. The only connection which we have so far been able to trace between these two levels of enquiry is the rule of twelve-side deficit, and this is evidently a very loose and tolerant form of restriction if we have to deal with reasonably high grades of N-hedron. If we are to display a model of say a 15-hedron, our freedom to choose particular n-gons as faces is not unlimited, but on the other hand it is perfectly obvious that there must be many ways in which we can exercise a decided preference or bias. We might decide, for instance, to introduce a 5-gon face at every opportunity, or we might take the opposite standpoint and refuse to use any 5-gon in our construction until absolutely compelled to do so. The imposition of such prejudices must make some difference to the final result, but how much difference? How many different kinds of 15-hedron are there, anyway? Unless we can form some impression of these matters we shall be unable to distinguish between features which merely exhibit the inescapable minimum of conformity to geometrical law and those which signal the existence of decision-making systems in the cells.

A comprehensive mathematical treatment would go far beyond our needs: it will be sufficient merely to enhance our intuitive perceptions by exploring a numerical example. At the left edge of a page set out in column the figure 6 three times and the figure 5 twelve times. This defines a possible combination of faces for a 15-hedron, because it satisfies the deficit rule. To make a different 15-hedron in the next column we can transfer sides at will from face to face leaving the total deficit unaffected. The only other limitation is that no face can be reduced below the grade of 2-gon. Thus in our second column we might have: one 7-gon, two 6-gons, eleven 5-gons and one 4-gon. Further operations of the same kind will enable us to generate other forms of 15-hedron, always with confidence that a model can be made whenever required. If at any stage we wish to proceed to examine 16-hedra instead we can simply add one 6-gon to any of our columns and work on with the necessary deficit intact. But if we mistakenly generate a column in which the deficit is no longer equal to 12, then we shall have specified a figure which cannot be constructed.

If we produce a few columns in this way and then begin to consider what would be involved in the actual manufacture of models for display, it will become apparent that some columns might be capable of more than one style of 3-dimensional embodiment. For example we have mentioned a 15-hedron with a 7-gon face and also a 4-gon. If we could make a model in which these faces were neighbours and another in which they were not, this would be a significant addition to our powers of geometrical invention.

A little experimentation along these lines will soon lead to the conviction that polyhedra are very diverse geometrical figures. For normal biological applications the more extreme variants can of course be safely disregarded: twelve 3-gons and three 14-gons is a real 15-hedron, but hardly an everyday shape for a cell. Even so, it is obviously necessary to prepare for the possibility that a tissue sample will yield such an array of different polyhedral forms that the observer's encounter with any single variety assumes the status of a rare accident which may never be repeated.

An example may be taken from Matzke (1948): it is chosen because the polyhedral grade of the cells was exceptionally low (about 10-hedral on average). Even in this simplified context, a sample of 200 cells yielded 36 polyhedral forms of which 14 were found only once and only 2 had frequencies exceeding 10% of sample. With higher polyhedral grades (and for any general scheme of tissue survey it would be necessary to go at least as far as 18-hedra) these statistical peculiarities are sharply accentuated. Thus Lier (1952) in a sample of 100 cells had 92 polyhedral forms of which only 7 were seen more than once.

Having regard to the requirements in microtechnique and statistical analysis, any attempt to deal with histological problems in terms of individual polyhedral figures is quite out of the question. Admittedly a very limited range of polyhedra attain frequencies which possibly might, if there were sufficient reason for doing so, be ascertained with some degree of accuracy. There is however no obvious indication at present that polyhedra which are specially frequent in tissues are of any particular importance or special interest in any other way. Conversely a polyhedral formulation is by no means assured of frequent occurrence simply because it approaches an average condition for the tissue. Marvin (1939b) established for a sample a set of (fractional) averages to which the nearest whole-number approximation was: (4 × 4-gons + 5 × 5-gons + 4 × 6-gons + 1 × 7-gon), which is a real 14-hedron. Marvin was never able to find a cell this shape, though he had several other 14-hedral types. To accomplish anything of a fundamental nature we obviously have to look behind this unsatisfying and essentially accidental class of information.

What has happened, in geometrical terms, is that the rule of twelve-side deficit, not surprisingly, has proved to be a weak and inadequate link between the scientific study of cells as N-hedra and the study of the faces of cells as n-gons. Complete independence we know there cannot be, but as a practical necessity, and until the technical resources at our disposal are greatly increased, our only proper course is to be prepared, wherever necessary, to dissociate the faces from their cells and pursue independent lines of enquiry.

4

The division of space

Our object is now to pursue a course of theoretical development as nearly as possible analogous to that followed in the early part of the previous chapter, but dealing with cells as solid bodies instead of flat outlines. That is to say, we aim to establish a statistical equilibrium in which the effect of mitosis is to recycle cells endlessly through a recognisable average course of geometric change. This time, however, we shall push the matter to conclusions and make a comprehensive comparison between our theoretical scheme and the results of laboratory observation.

The general strategy, given the experience of the earlier work, is obvious enough. We are to postulate a steady state, calculate the required life-course of a cell in terms of polyhedral upgrading, and so obtain an equation which will relate the polyhedral state of a cell to its volume or (which in a steady-state tissue amounts to the same thing) its mitotic phase. After we have obtained such an equation it must be subjected to verification by measurement of specimens. Then, if the comparison proves satisfactory, it will be time for us to calculate (if we can) the corresponding cycle of changes affecting the polygonal faces of the cells, and similarly submit that calculation to the test of observation.

We begin by noting that in moving from the surface problem to the space problem we have an orderly extension of the basic laws of construction:

Surface divided into polygons
  (*a*) Two sides meeting at a wall.
  (*b*) Three cells and three walls meeting at a corner.

Space divided into polyhedra
  (*a*) Two faces meeting at a wall.
  (*b*) Three cells and three walls meeting at an edge.
  (*c*) Four cells, six walls and four edges meeting at a vertex.

For the surface problem, these basic specifications were all that we needed to prove an exact average number of sides per cell. Can we prove a corresponding average number of faces for cells considered as solid figures?

The problem of intrusive growth

We consider Fig. 23a, which is the solid analogue of Fig. 16a at p. 38. Two cells A and B face each other at opposite ends of an intervening edge. Surrounding the edge are three cells C, D and E, of which E is deemed to underlie both C and D. From a view-point in A, one would see the arrangement shown in Fig. 23b. Evidently it would be possible for A and

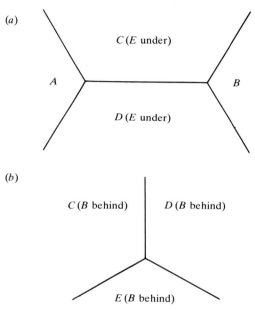

Fig. 23. Arrangement of five cells in a solid tissue. (a) is the view of an edge, and should be compared with Fig. 16a at p. 38, but there is here another cell underneath. (b) is a view from A to B. It is possible for A and B to achieve a meeting by a triangular tunnel without causing any loss of contact between C D E.

B to establish a mutual 3-gon interface by opening out into a triangular tunnel the edge which appears as a horizontal line in Fig. 23a and as a central point in Fig. 23b. A move of this kind, which we may call intrusive growth, adds a new cell-to-cell contact without destroying any of those which existed before.

There is consequently no principle of conservation such as was available in the corresponding two-dimensional enquiry at p. 37. So long as there is any possibility of intrusive growth, the average number of faces per cell is mathematically indeterminate, or to be more accurate we may perhaps be able to fix a minimum polyhedral grade for cells but must be prepared to find in real specimens that intrusive adjustments have produced extra cell faces, raising the observed average unpredictably above any minimum which can be derived from abstract principles. The study of solid tissues is therefore of a different philosophical nature from that of tissue sections. That mitotic mother-cells in a section of steady-state tissue must average seven sides each is a geometrical certainty requiring no observational test: an equivalent generalisation about mitosis in the solid is attainable only

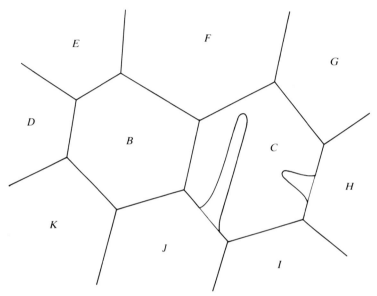

Fig. 24. Development of additional faces by intrusive growth. All the cells shown are viewed from another cell *A* which overlies them. *J* is producing a tubular outgrowth across the *A/C* interface and will shortly establish contact with *F*, and *H* may later extend to touch *B*. By *selective* intrusion the number of interfaces in a solid tissue can theoretically increase without limit, but in reality the level of such activity is generally very low.

by making some suitably chosen assumption which must then stand or fall according to the correspondence (or lack of correspondence) between theory and observation.

It might at first appear that there could be some upper natural limit to the generation of new cell faces by intrusive re-arrangement. If we suppose intrusion to occur at every possible point, will not the face-producing potentialities of the process eventually be exhausted? This question must be answered in the negative: if we allow unrestricted intrusion the number of faces per cell goes on increasing indefinitely. For consider Fig. 24, in which two polygons represent two faces between a cell *A* and two of its neighbours *B* and *C*. Other surrounding cells have been lettered *D*, *E*, etc. From *J*, a glove-finger process is intruding into the interface between *A* and *C* and will soon extend to touch *F*. A similar process from *H* will afterwards be able to reach *B* by passing over or under the extension of *J*. Having completed its crossing of the *A/C* interface, the tube from *H* will be free to continue its course: it might go right across the *A/B* interface to touch *D*, and this would not prevent subsequent contact between *K* and *F* or between *E* and *J*. To the scope of such developments there may be some biological limit. Geometrical limit there certainly is not, and we cannot proceed further until this deficiency is made good.

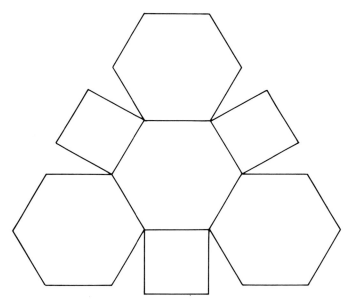

Fig. 25. Assembly of squares and hexagons to make one half
of an orthic tetrakaidecahedron.

The Kelvin proofs

It is a fact of observation that in apparently homogeneous tissues the average number of faces per cell differs only fractionally from fourteen. For many years in the biological literature this number has been associated with a mathematical proof by Kelvin (1887), here distinguished as Kelvin's *first* proof.

The attraction of Kelvin's first proof for biologists has been twofold. Firstly the proof leads to a tetrakaidecahedron, or fourteen-faced solid, and the coincidence of number very naturally makes an impression upon the mind. Secondly, the problem which Kelvin set himself was that of partitioning space in the way which would give the absolute minimum of wall area per unit of enclosed volume. Philosophically therefore the matter is related to surface tension considerations, or to a compression model in which the original spherical form of the units is modified only to the

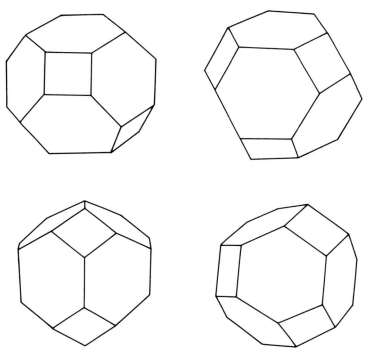

Fig. 26. Perspective views of orthic tetrakaidecahedra. By rotating the page these diagrams can be made to reproduce most of the characteristic appearances to which this solid can give rise. In combining such blocks into a continuous stack, hexagonal faces must be juxtaposed so that the square faces of one cell alternate with those of the other. In the solid, therefore, a hexagonal interface has a square attached to every one of its six edges (but a square meets only hexagons, and no other square).

unavoidably minimum extent. To obtain an approximation of Kelvin's 14-hedron see Fig. 25, in which a regular hexagon is surrounded by hexagons and squares alternately. Cut this out of card and fold up the peripheral figures until their edges meet, and the result is a kind of angular bowl. Two such bowls will combine into what is known as an orthic tetrakaidecahedron (Fig. 26). The Kelvin 14-hedron differs from this in a slight curvature of all its edges; in this final adjustment the 4-gon faces

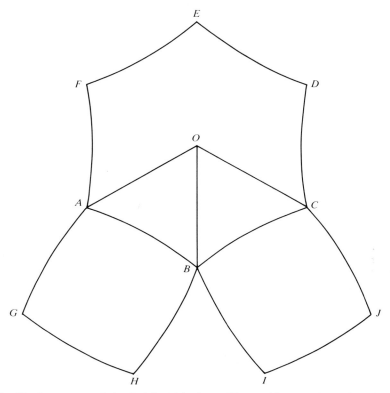

Fig. 27. Curvatures of the Kelvin 14-hedron. We consider one 6-gon (*ABCDEF*) and the 4-gons adjoining two of its sides (*ABHG* and *BCJI*). The 4-gons are to be treated as flat and rigid doors hinged on the straight lines *AB* and *BC*, while the 6-gon is to behave like a liquid film. We raise *GH* above the paper, causing the curved line *AB* to dip a little *below* the page. *OAB* thus becomes a depressed area. *IJ* must be pushed downwards (see caption to Fig. 26), and *OBC* will then rise into a low ridge. Continuing round the 6-gon we shall raise three of its sectors and depress the other three. The lines *OA*, *OB* etc. will remain in the plane of the paper, the *total* curvature of the 6-gon will remain everywhere zero, and the concavity of the 6-gon edges will be, from our viewpoint, about halved. The curves in the diagram are almost, but not quite, circular arcs.

remain flat, though ceasing to be perfect squares, while the 6-gons are thrown into gentle saddle-type undulations (Fig. 27).

The Kelvin 14-hedron has been the subject of an active propaganda, and many biologists have been conditioned to expect this particular polyhedral form to figure largely in histological studies. Such expectations are, and always have been, totally unreasonable and unfounded. The conclusion must be that Kelvin's paper has been cited as an authority more often than it has been properly read and understood.

Kelvin was able to solve his problem only by making two assumptions, neither of which is compatible with the existence of life in the system. His first assumption was quite explicit: space was to be filled with 'equal and similar' polyhedra. Biologically this requires total uniformity of cell volume throughout eternity and to an infinite distance in every direction. Growth, mitosis, individual variation of cells are all equally forbidden. The willingness of so many biologists to overlook so much for so long is remarkable. Kelvin's second assumption is no less discouraging from the biological standpoint. It is not expressly stated in his text but is implied in the degree of abstraction and idealisation which is normal and proper to an argument in pure mathematics. The assumption was simply that there was no need to consider the practical aspects of manufacture: the space-filling array of 14-hedra could be deemed to spring into existence in all its perfection at a given moment. For an idea in the mind of a mathematician this is a legitimate presumption. For a system which is actually to be displayed in material form, it is not. It is tantamount to the presumption that any conceivable form of building could be erected without scaffolding, which is untrue, for example, for a masonry arch.

Nothing is more revealing than the fact that Kelvin, a brilliant and tireless experimenter, having proved the ideal shape for close-packed equal-sized soap-bubbles, and having exhausted his ingenuity in the laboratory, was obliged to resort to the soldering-iron. His 14-hedron was ultimately produced by the aid of a complete wire skeleton in which the position of every edge was *forced* upon the liquid films. It has since proved possible (Dodd, 1955) to produce a Kelvin bubble in a slightly less unnatural manner, but nothing can disguise the fact that the Kelvin 14-hedron for all practical purposes does not arise in tissues or compression models. For example Macior & Matzke (1951) in a tissue sample found that even the Kelvin combination of faces (eight 6-gons and six 4-gons) occurred in only about one cell in fifty. All material systems abound in 5-gon faces, which on Kelvin's first proof should not occur at all.

These grave discrepancies between theory and observation evidently have nothing to do with the question of standardisation of size. Systems of equal bubbles conform to the Kelvin 14-hedron little if any more convincingly than do systems of unequal bubbles. The trouble lies almost entirely in the mechanics of construction. Kelvin's premises amount to the statement that everything is to depend on the inclination of each cell to dissipate energy by reducing its surface, for which purpose the energy-level of the sphere must be reckoned as zero. But we have already seen (p. 14) that any reasonably compact polyhedron such as can be expected to occur in a compression model is already near to that level. The release of energy which is available to change one 14-hedron into another can only be very small. Furthermore, as between any two geometric figures, a physical transformation will only take place if all the intermediate conditions which have to be passed through can be arranged in a consistent gradient which gives a release of energy all the way. A river cannot climb a hill to reach low ground on the other side, nor can a tissue make temporarily disadvantageous adjustments on the way to some more distant benefit.

The correct view is therefore a statistical one. If we distort a foam, bubbles are pushed over bubbles and drop into successive positions of stability. During this process events in different parts of the system are uncoordinated and unsynchronised. When geometrical perfection is displayed at some point A, conditions will generally be less than perfect at some other point B. If we move B, the system will be disturbed, and the ideal configuration at A is likely to be destroyed if we have no means of holding it in place during the reorganisation. This is simply the scaffolding requirement in a new form. There is absolutely nothing wrong with the Kelvin geometry, except that without some kind of support, or an army of microscopic intelligent assistants, the laws of chance will prevent us from ever getting all the pieces into position at one and the same time. We know that if we could erect it, it would stand, but we also know that we cannot erect it.

From the biological point of view the only aspect of Kelvin's first proof which is genuinely of interest is the number 14. But this in fact is not mathematically related either to surface-tension principles or to any particular shape of cell. To appreciate this we refer to the astonishing *second* proof (Kelvin, 1894), which is mathematically far more profound, and of which the first is only a rather unimportant special case.

The basis of the second proof is simply that space is to be divided into compartments which are equal in size, shape, and orientation. If these requirements are satisfied, each cell is in contact with exactly fourteen neighbours, no more, no less. Shape has nothing to do with it, so long as space is entirely filled, and the rules of uniformity are not infringed. If we desire, for example, to fill space with identical cells each of which has eighteen faces this can be done, but in designing our cells we shall find it necessary to make them wedge-shaped and stack them with their thin ends right and left alternately, which is a breach of Kelvin's stipulation about uniform orientation.

To obtain a full perception of these truths is not easy, nor for biological purposes is it necessary to press the matter very far. It may be useful to examine Fig. 28, a cell-form derived from a system of hexagonal columns in which there are transverse partitions suitably staggered so that each of the six side-walls of the cell consists of upper and lower panels. With the hexagons at top and bottom this makes fourteen faces in all. Cells of this form can obviously be designed so as to exemplify the conditions of the second proof.

Starting from this or any other regular arrangement we may allow the cell to exhibit intrusive behaviour in any style we care to adopt. The stipulation which leads into Kelvin's second proof may then be expressed by saying that any intrusive activity which is permitted to one cell must

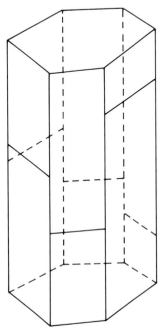

Fig. 28. Derivation of a fourteen-faced cell form from hexagonal columns with transverse partitions. Each cell has top and bottom, plus six sides each of which is divided into upper and lower panels by the (staggered) edges of the partitions in adjoining columns.

be duplicated exactly in every other cell: intrusive development is to be standardised in shape, size, timing, and direction. The rule of fourteen contacts will then be inviolable. To show that this is true for some geometrically simple particular form of intrusive adjustment can require only care and patience. Kelvin's achievement was to show that it is true universally.

From the biological standpoint the second proof is of more interest than the first, because it has the advantage of giving us a pure theoretical basis for the number 14, which is closely related to observational experience, without involving us (as the first proof would be liable to do) in unrealistic expectations about the polygonal form of cell faces. It will have been noted in connection with Fig. 24 (p. 56) that any highly developed system of intrusive growth produces appearances altogether foreign to the common experiences of histological enquiry. If we assume a tissue to have no more than a low level of intrusive activity, and if we further assume an approximate uniformity of cell size, then the principles of the second Kelvin proof will generate the expectation that the average number of faces per cell may be slightly more than fourteen, but not very much more.

There is here an obvious point of weakness, in that uniformity of cell volume is an unrealistic expectation and we have not considered the effects of mixing large and small cells. Potentially there would appear to be a danger that variability of size might disturb our conclusion about the average polyhedral grade, and that the number of faces per cell might in some circumstances even be less than fourteen. For the time being we have no convenient way of disposing of this problem, but a partial explanation will be given at p. 89. On the other hand we have gained considerably by freeing ourselves from any too-rigid connection between the polyhedral and polygonal levels of analysis. We may proceed at once to establish a cycle of change of polyhedral form, leaving the shapes of cell faces to be treated independently in another chapter.

A working hypothesis for mitosis in polyhedral cells

Let us provisionally assume that:

(*a*) new cell faces arise in a tissue only by mitotic partition of existing polyhedra;

(*b*) the average mitotic event is equivalent to the transverse section of a hexagonal prism.

Of these two stipulations (*a*) is simply a restatement of the idea that intrusive activity is unlikely to be a conspicuous feature of undifferentiated tissue and can be neglected in the early stages of analysis. The supposition (*b*) requires rather more attention.

We are to divide a hollow N-hedron by a partition across its cavity. The partition will itself be an n-gon, the sides of which will divide into two each of n faces of the N-hedron. The other faces of the N-hedron (i.e. those not intersected by the partition) remain unchanged, but just as we found in the analogous case in two dimensions (p. 39) there is an injection of n new faces into the surrounding interphase tissue. We are proposing that the average value of n should be 6 exactly. The partition formed in mitosis being a 6-gon, we have the following generalisations:

(a) Mitosis in a mother-cell which is an M-hedron results on average in a total endowment for the daughter-cells of $(M+8)$ faces.

(b) A mitotic division imparts on average 6 additional faces to cells not involved in the mitosis.

These are in principle parallel to the statements at p. 39 and will similarly imply a circulatory pattern of geometric change, and lead us almost immediately to an equation for steady-state equilibrium.

We may pause however to dispose of an apparent contradiction. We showed at p. 41 that the average section of a mitotic mother-cell in steady-state tissue must be a 7-gon, yet here we propose to make the partition formed in that cell a 6-gon. How are these figures to be reconciled? In reality there is no problem here, because the numbers relate to examination of a cell in two mutually perpendicular lines of sight. We may view a cell from the east and see the mitotic partition from its edge, as a line dividing a 7-gon. Moving round to view the same cell from the north we see the partition in face view and appreciate that it is itself a 6-gon. There is no real difficulty in such a concept and we retain the notion of randomised states as the base level from which structural differentiation must be considered to begin. The *least organised* assumption that is available to us is that the mitotic partition shall be any random section through any randomly chosen cell, and this leads at once to the 6-gon average for the partition itself and also to the 14-hedral average for cells.

The mitotic cycle for polyhedral forms

Following the pattern already familiar from p. 41 we have as a condition of equilibrium

$$\tfrac{1}{3}\{M+(m+8)\} = 14$$

and $M = 17$. In average terms only, therefore, we find that a cell is born as a $12\tfrac{1}{2}$-hedron, and divides as a 17-hedron, acquiring $4\tfrac{1}{2}$ faces during interphase because of the mitotic activity of its neighbours. Direct observation of these characteristic numbers of the mitotic cycle cannot be very accurate, but Matzke & Duffy (1956) obtained 12.61, 16.84, 4.23.

In order to make any practical use of this we shall need to establish a relationship with the volume of the cell. Evidently it is only relative volume with which we shall be concerned. The geometrical status of a cell is independent of absolute size but depends on whether the cell is larger or smaller than its neighbours. We define

$$V = \text{specific volume} = \left|\frac{\text{volume of any selected cell}}{\text{average volume of all cells}}\right|,$$

and if we retain N for the number of faces of a polyhedron we require for the analysis of observational data an equation relating V and N.

We already possess three points on the graph of the required expression. We have $(V = \frac{3}{4}, N = 12\frac{1}{2})$, $(V = 1, N = 14)$, and $(V = \frac{3}{2}, N = 17)$. These points are collinear, and fit the equation

$$N = 14 + 6(V - 1), \quad \text{or} \quad V = \frac{1}{6}(N - 8)$$

so if we are to justify our working to this point we have to demonstrate the conformity of observations to this equation at least for values of N which are not too far removed from 14. The test is naturally to be conducted upon the level of general averages: the validity of a statistical principle cannot be endangered by erratic behaviour in a particular cell.

The limits of extrapolation

V being the volume of a cell, and N being an attribute merely of the outer surface of that cell, it must be disturbing to any truly philosophical mind to be confronted with a linear relationship between the two. Linearity, one might think, would be more likely to exist between the cube root of the volume and the square root of the surface. This reflection prompts us to examine more carefully the implications of very large or very small values of V and N.

Consider, say, a 100-hedron, which is not such a far-fetched example as might appear, for we shall shortly have to review observational data relating to the 47-hedron. In randomised tissue, the neighbours of any very high polyhedron will constitute a reasonably large random sample of the population. Their mean specific volume consequently approximates to 1. The volume we are to attribute to a 100-hedron from the equation above is about 15, and the ratio of diameters therefore about $2\frac{1}{2}:1$. Simply by making this ratio the basis of a diagram and sketching the juxtaposition of an isolated large cell with its smaller neighbours, one very soon arrives at the conviction that the larger cell will approximate rather closely to a sphere, that the small cells will be significantly flattened against the side of the large one, and that even an enormous further enlargement of the giant cell would not very greatly increase the flattening effect which is already apparent. There is, in short, a natural limit to the size of face which can exist between cells very disparate in size. This limit can never be reached with cells of finite size, but for $N = 100$ the approach to the limit is already rather close. Extension of the argument leads to the infinitely large cell which is a perfect sphere with its surface area simply proportional to N. This gives the equation of a curve:

$$N = KV^{\frac{2}{3}}$$

where K is a constant. We have, therefore, now two limiting equations which we may call the linear and spherical predictions respectively, and our revised theoretical expectation must be that as N increases to very high values, the observational data will progressively forsake the straight line and take on the curvature of the other equation. To carry out this graphical operation requires results from the next chapter, but the general effect can be seen by referring forward to Fig. 30 at p. 80.

At the other end of the range, the application of our algebra must be terminated at some point by the geometrical peculiarities of the very low polyhedra. Once a cell has shrunk to such a size as to become a 4-hedron, there ceases to be any basis for consistent mathematical treatment of any kind, because a 4-hedron can simply collapse into a single vertex of the system and disappear completely without further loss of faces. Allowing curvature of the faces it is quite possible to visualise cells in the form of a 2-hedron or 3-hedron, but if such cells were to exist we would be obliged to consider them as a separate class of phenomenon: they manifestly cannot be fitted into the general scheme of tissue structure with which we are currently concerned.

The minimum range of polyhedral diversity which is necessary to maintain a balanced mitotic circulation of forms extends from the 12-hedron to the 17-hedron inclusive. As a rough allowance for individually deviant cells let us double this range to include the 9-hedron and 20-hedron as virtual extremes. As a matter of observational experience, this is roughly the range encountered in unspecialised tissues.

Standards of observational accuracy

Taking the approximate range of polyhedral forms just defined, it is easy to deduce a rough approximation for the standard error of N in a sample of unselected cells from a steady-state tissue. We may reasonably take the standard deviation for single cells as about ± 2 faces per cell. For standard error of a sample we divide by the square root of sample size, and reasonable confidence limits will be $\pm$ twice the standard error. We have therefore from a sample of 100 cells (a common form of practical investigation) a range of uncertainty of about ± 0.4 of a face on the general 14-hedral average. To establish 14 ± 0.1 with any normal standard of statistical justification is unlikely to require less than a thousand-cell sample. Nothing of the kind has ever been attempted. To form a clear impression of the practical limits of enquiry is important for us. It may be helpful to point out that ± 0.1 of a face is about $\pm 1\frac{1}{2}\%$ in volume, and consequently $\pm \frac{1}{2}\%$ in linear dimensions of the cell. In biological investigations an accuracy of $\pm \frac{1}{2}\%$ is reached but seldom, and we need not pay attention to the various small decimal discrepancies which have been reported for the average number of faces per cell in tissue samples.

Table 1. *Observational test of the equation $N = 14+6(V-1)$. Data for pith cells (C), bubbles (B), and compressed shot (S)*

V	Calculated N	Observation	Difference	Mean differences
0.30	9.80	9.50 (S)	−0.30	
0.30	9.80	9.68 (B)	−0.12	
0.31	9.86	10.70 (C)	+0.84	+0.02
0.37	10.22	9.50 (C)	−0.72	
0.55	11.30	11.70 (C)	+0.40	
0.56	11.36	11.90 (B)	+0.54	
0.56	11.36	12.04 (S)	+0.68	
0.60	11.60	11.64 (C)	+0.04	+0.29
0.62	11.72	11.70 (C)	−0.02	
0.83	12.98	13.20 (B)	+0.22	
0.83	12.98	13.30 (S)	+0.32	
0.91	13.46	13.60 (C)	+0.14	
0.96	13.76	13.53 (C)	−0.23	−0.13
1.28	15.68	15.60 (C)	−0.08	
1.40	16.40	15.61 (C)	−0.79	
1.52	17.12	15.23 (C)	−1.89	
1.64	17.84	17.53 (C)	−0.31	
2.03	20.18	18.60 (C)	−1.58	−1.64
2.40	22.40	19.98 (S)	−2.42	
2.40	22.40	20.42 (B)	−1.98	
4.50	35.00	25.62 (S)	−9.38	−8.52
4.50	35.00	27.34 (B)	−7.66	
6.60	47.60	30.06 (B)	−17.56	−17.48
6.60	47.60	30.18 (S)	−17.42	

Analysis of data

Within the readily attainable standards of accuracy the general average number of faces can be treated as 14, on the basis of numerous published statements. This is true for cells and also for models such as foams and compressed shot. It is true whether the units are of standardised volume or assorted. We have therefore no lack of confirmation of our central graph-point ($V = 1$, $N = 14$).

For other points we are dependent upon a more restricted range of published determinations. For plant cells we use the data of Marvin (1939*b*, 1944) and Lewis (1944), for shot those of Matzke (1939), and for bubbles those of Matzke & Nestler (1946).

Quite apart from the difference in physical composition of these systems, they exhibit substantial differences in geometrical conformation. The cells were extracted from natural tissue in which there was presumably a continuous gradation of size, just as we have assumed throughout the course of our theoretical development. The shot and bubble experiments, however, were highly artificial in that the units employed were of two sizes only. Matzke mixed, in varying proportions, large and small shot with diameters in a ratio of 2:1 and volumes consequently in a ratio of 8:1. The absolute volume of a shot in the whole series of experiments had only two possible values: its specific volume, as we have defined it, was more variable only because the composition of the mixture could be adjusted to give different *average* volumes. None of the original observers employed the concept of specific volume at all, so a certain amount of recalculation is necessary to place all the results upon a common basis.

When this is done all three sorts of observation (for cells, bubbles and shot) are found to run beautifully together in a single sequence (Table 1). In view of the great difference of population structure we could hardly have ventured to forecast this degree of uniformity.

From the column of mean differences it appears that at least from the 10-hedron to the 15-hedron the relationship between shape and volume is in almost perfect agreement with the simple linear equation which arises from the hypothesis of randomised mitosis. From about the 16-hedron upwards there is the expected increasing correction to be applied: this we can understand in broad descriptive terms, though not yet upon a strictly quantitative basis, in terms of the inescapable flattening of any small cell which is juxtaposed to a very much larger neighbour.

Variability of polyhedral grade

We have at several points recognised that the geometry of a cell cannot be completely determined by the stage which that cell has reached in the mitotic cycle. Quite apart from the special problem presented by the occurrence of fractional averages, some room must be left for the operation of purely accidental forces. Some arguments have been based on the provisional guess that variation due to accident might roughly double the variation imposed upon the tissue by necessities of mitotic balance. The matter is open to observational enquiry and the data employed in the preparation of Table 1 will yield practical orders of magnitude for the accidental variability. As a working approximation from actual measurements we may say:

(*a*) For specific volumes up to, and somewhat exceeding, unity, up to 33% of individual cells may have the whole number of faces which is nearest to any theoretical prediction, while 90% of individuals will be within a range of ± 2 faces.

(*b*) For large specific volumes (of 3 to 5 or thereabouts) there is greater individual variability. About 90% of cells will be within ±3 faces of the predicted value, but no particular face-count is likely to occur in much more than 15% of cells.

Theoretically a specific volume which yields a predicted face-count which is integral, or nearly so, ought to produce a narrower statistical spread than one which falls midway between two whole numbers. In reality the published observations are insufficient in number or quality to permit such an enquiry.

These investigations are of value primarily in roughly confirming the original guess. The range of polyhedral grades which is absolutely necessary for the expression of a mitotic circulation of forms runs from 12 to 17 inclusive, and we shall cover virtually all the cell-forms in undifferentiated tissues by adding 3 grades at each end (9-hedron to 20-hedron, as suggested at p. 66).

Difficulties of volumetric measurement

To measure the volume of an individual cell is a necessary step in establishing a firm relationship between volume and shape. Once a calibration has been found, however, repeated volumetric measurement becomes an unattractive proposition in the laboratory. It is an extravagant use of observer-time, being very much slower than a count of faces. A face-count would not be an acceptable substitute for the measured volume of an individual cell, because the margin of uncertainty would be very large. We are forced by observation and by reasonable *a priori* expectation alike to admit such a measure of individual variability that for example a 14-hedron might stand at absolutely any position in the mitotic cycle. For single cells the designation of a 14-hedron does not signify a particular volume but rather a range of volumetric fluctuation which probably exceeds a ratio of 2:1. This in no way invalidates the conclusion that where reasonably large samples or subsamples are concerned the best tactics will be to base the work on face-counts and to convert these to volumes, if desired, by using a standard calibration prepared in advance once and for all.

From this point of view the calibration offered in Table 1 is manifestly not all that could be wished, but if suitably presented in a smooth graphical form it will constitute a serviceable laboratory tool until better data become available. The existence of even an imperfect calibration at this point is important, because it means that the studies to be pursued in later chapters can be carried on without the burden of direct volumetric measurement.

Polyhedral cells at a surface

The course of theoretical development which we have followed to this point
has depended heavily upon the two related mathematical concepts of
randomness and symmetry. Buried deep inside a continuous mass of tissue,
all directions have been the same to us, no journey we could undertake
would ever bring us to any place where conditions were more than
transitorily and accidentally different from conditions anywhere else. As
soon as we consider cells at or near the outer surface of a tissue mass this
helpful simplicity is lost, and mathematical difficulties of a higher order
automatically arise. It does not appear that we can have any immediate
prospect of finding any comprehensive general solution to the problems
of tissue construction near to an outer surface.

Suppose that we cut off a steady-state solid tissue by a plane of section
through its interior. We thereby expose cell cavities upon a plane which
is now to become an external surface. We may close up all the opened cells
by supplying a suitable outer wall and cementing this to the edges left by
the knife. At this stage we may say:

(*a*) The average polyhedral grade of the surface cells is that of an
11-hedron, because although the knife will strike cells in various ways its
average impact will be equivalent to the equal division of a 14-hedron.

(*b*) The average polygonal grade of the surface outcrop of a cell must
be 6 exactly, from our proof at p. 37.

(*c*) The average specific volume of a surface cell will be $\frac{1}{2}$.

Of these results (*a*) and (*b*) need cause no special concern, but (*c*) implies
a very unsatisfactory state of affairs.

Sectioning a randomised tissue in the manner postulated will mean
that the size of cell in the new surface layer will be extraordinarily variable.
In some cases a cell will be only slightly grazed by the knife and survive
in the changed situation with its volume scarcely diminished at all. At the
other extreme a cell may come through the operation as a mere fragment.
Plane section of randomised tissue consequently does not generate a
biologically plausible form of geometrical construction. It is easy to see that
the model might be somewhat improved by a simple plastic deformation.
Let the flat outer wall be drawn outwards a little so that the cells in the
surface layer are stretched perpendicular to the surface and restored to a
mean specific volume of unity or thereabouts. This will conform to
ordinary biological reality to the extent that in surface cells the walls which
run out to the exterior must, upon surface-tension principles, meet the
surface at right angles or nearly so. The form of a surface cell, in fact, must
be hybrid: its inner end is a normal compression-model polyhedron
approximating to a sphere, but its outer end must be substantially a prism

with parallel sides and cut off square against the tissue boundary. Our proposed plastic deformation, stretching the cells outwards, is simply a device for introducing into our model the necessary prismatic component of cell shape.

Even after secondary adjustment, however, a cut through randomised tissue will not afford a satisfactory representation of real surface structures as they are observed in histological work. We have the incurable difficulty of excessively variable polyhedral grade, no matter what may be done by plastic deformation to equalise the cells in other respects.

Real tissue surfaces always involve a conspicuous alignment of cells. Within a certain distance from the surface we must recognise a zone in which it is not permitted that the centre of gravity of any cell should occur. To put the matter in another way, every cell which touches the outer surface at all will normally display in that surface very nearly its maximum area of cross-section. It is consequently impossible to obtain an adequate model by a single slice through randomised tissue because the vital element of stratification is lacking. We can only build up a credible collection of surface cell-forms by sectioning polyhedra selectively, and the moment we adopt a selective basis we lose the only obvious source from which it would be possible to draw firm predictions about polyhedral grades.

Observations tend to show that surface cells which are not strikingly disparate in size and proportions from the cells which lie below them will have an average polyhedral grade not far removed from 11. By considering the limiting case in which small surface cells overlie a single enormous internal cell it is easy to see that as the size of the outer cells diminishes relative to that of the inner ones their average polyhedral grade must fall towards a minimum attainable level of 8. Disproportion in the other direction will evidently lead to unlimited increase, with every surface cell possessing interfaces with a multitude of tiny internal cells.

The general nature of the balance is therefore reasonably clear, but only in broad descriptive terms. A quantitative theoretical treatment is lacking. It is, of course, quite possible to prove that certain geometrical consequences must ensue from some specified particular configuration. We can, if we wish, define an arrangement of cells which will give a layer such as a plant epidermis a mean polyhedral grade such as 11 or 10 exactly, at our own choice. All such exercises however are subject to the same biological objections as Kelvin's first proof. They only succeed at all because they assume a uniformity of size and a regularity of arrangement which no living organism can reasonably be expected to approach. The only treatment which would be of general application in biology would be one which extracted statistical regularity from a chaotic situation *without* needing to insist on artificial standardisation at the commencement of the work. It does not seem that any real illumination of the surface problem can be expected for the time being. For example, we cannot establish an equation of mitotic equilibrium in surface cells because there is no acceptable basis of assumption about the shape of the mitotic partition. If a cell at the surface divides by a partition perpendicular to the surface there is a certain natural impulse to suppose that such partitions might on average take the form of a 5-gon, and observations might tend to support or contradict that idea. What we lack at the moment is any clear theoretical reason for preferring any exact number rather than any other. Anything in the range $4\frac{1}{2}$–$5\frac{1}{2}$ seems equally likely to be right.

Because surface cells tend to be conveniently accessible to observation they are often chosen as observational subjects, and in general there may be no need to criticise this procedure. Any supposition, however, that surface cells exist in a simplified geometrical context, or that their smaller number of faces will necessarily make their three-dimensional shape easier to understand, is almost certainly unsound. So far as any theoretical explanation is involved, the study of surface cells offers no 'soft option' in tissue geometry.

5

The faces of cells

We have now to consider the polygonal faces of cells, in the first instance as independent entities which can be treated separately from the particular polyhedral forms to which they happen to be attached. Just as a tissue can be statistically analysed as a population of cells, so it can be analysed as a population of faces.

The style of analysis must obviously be somewhat different from that adopted in previous chapters, because we no longer have a simple one-to-one relationship between the objects of our mathematical study on the one hand and the mitotic cycle on the other. All cells arise by division of pre-existing cells but cell faces have two modes of origin, not one. Some faces are formed by division of pre-existing faces, but others are newly created in the mitotic partition itself. The geometry of cell faces, therefore, unlike the geometry of polyhedral cells or polygonal cell sections, is only *partially* circulatory. When a cell divides by a partition, the new areas of cell surface which are created in that partition exemplify conditions to which there can never be any return. We have a population of polygons in which a large element of circulation occurs, but instead of being a closed circulation we have to think of it as being fed by a steady influx of new material which enters along an irreversible pathway. This will make our calculations more complicated, but the difficulty is not great enough to prevent us from proceeding.

Sides, edges, and vertices

It will be convenient to have for reference those characteristics of cell surfaces which follow directly from the study of polyhedral figures. In Table 2 we have the cell-forms from 9-hedron to 20-hedron inclusive, with examples of higher figures, up to the 50-hedron. An explanation of the last column must be postponed until pp. 76–8, but the remaining entries are calculated very simply in a manner which can be illustrated for the 14-hedron as follows:

A 14-hedron has 14 faces with a total of $(14 \times 6 - 12) = 72$ sides. Every edge involves a junction of two faces and therefore incorporates two sides, so

Table 2. *Characteristics of polyhedral figures of cells*

N	Edges	Vertices	Average n	Average f
9	21	14	4.667	0.58
10	24	16	4.800	0.70
11	27	18	4.909	0.80
12	30	20	5.000	0.87
13	33	22	5.077	0.94
14	36	24	5.143	1.00
15	39	26	5.200	1.05
16	42	28	5.250	1.10
17	45	30	5.294	1.14
18	48	32	5.333	1.17
19	51	34	5.368	1.20
20	54	36	5.400	1.22
30	84	56	5.600	1.40
40	114	76	5.700	1.48
50	144	96	5.760	1.53

there must be 36 edges. Three edges are required to run into each vertex, but as every edge is double-ended the number of vertices is 24, not 12. The average polygonal grade of all the faces is $72 \div 14 = 5.143$.

The origins of new cell faces

We have to envisage the mitotic division of a cell in terms of the production of a partition wall which on average will be a 6-gon. We are not now concerned any longer with the fact that such division is related to a 17-hedral form of cell. All we need to consider is the effect of a newly formed 6-gon partition upon pre-existing cell faces in its vicinity. In Fig. 29a we see such a partition in surface view, while Fig. 29b shows an edge view.

The edges of the partition have intersected six faces of the mitotic mother cell and also six faces of neighbour cells, that is twelve faces in all. Every one of these twelve faces has been divided (indicated by the letter D in Fig. 29a). The partition itself constitutes two newly created faces and is labelled 2C (i.e. two created). We have therefore a total production of new faces of fourteen, twelve by division and two by creation. In our equations, therefore, we must always reckon the newly created faces as $\frac{1}{7}$ of total output, or $\frac{1}{6}$ of the number arising by division. To express the matter in a different way, if faces arose only by division of existing faces the division of faces would have to run at fourteen times the mitotic rate: in fact it operates at only twelve times mitotic rate, the deficiency being made good by new creation.

Turning now to the question of the sides of faces, it will be evident that any face which is not itself being divided, but which shares an edge which is crossed by the anchorage line of a new partition, will be upgraded by one side. The faces which are upgraded in Fig. 29a are marked with a U, which occurs twelve times. Division of a face is subject to the normal rule

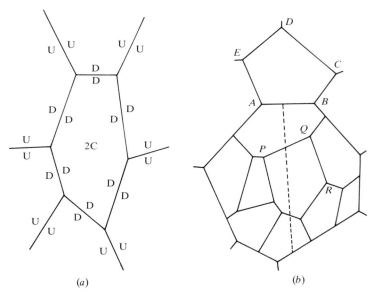

(a)   (b)

Fig. 29. Geometrical consequences of mitosis. (a) A 6-gon mitotic partition in surface view represents two newly created cell faces (2C). Its effects on neighbouring faces involve events of face-division (D) and of polygonal upgrading (U). (b) A mitotic partition in edge view (broken line). The cell being divided presents to our view nine faces (about right for a 17-hedron) of which three are undergoing division. *ABCDE* is an interface between two neighbours of the dividing cell, and is being upgraded from 5-gon to 6-gon because the edge of the partition intersects *AB*. The external interface which stands on *PQ* will be similarly upgraded, but that which stands on *QR* will remain unchanged. A 17-hedron has 45 external interfaces standing upon it, and a 6-gon partition will upgrade only 6 of these.

that the two daughters of any polygon jointly inherit all their mother's sides, plus four. The total addition of sides to the tissue is from Fig. 29a, therefore:

12 faces divided, 4 additional sides per parent face = 48
12 faces upgraded 1 side each   = 12
2 newly created 6-gons   = 12
                                                       ——
          Total   72

This is precisely what is needed for a steady state, as (14 faces + 72 sides) is the constitution of a 14-hedron, which we know to be the average cell-form.

An equation of equilibrium for cell faces

Our next objective must be to establish a standard cycle of change for cell faces just as we did at p. 41 for cell sections and at p. 64 for cells as polyhedra. For this purpose it will be convenient to take as our unitary transaction not the division of a cell but the division of a face, or rather the simultaneous division of two faces, for we have no need to separate the face of the mitotic mother cell from the face of its interphase neighbour. Let us denote as an *m*-gon a polygon which is to be crossed by the anchorage line of a mitotic partition, and let us then seek an average value of *m* which will secure the permanence of the steady state.

First, as to faces, our chosen *m*-gon constitutes two faces which will divide into four. But this account is incomplete: we have to credit the proposed transaction with its due share of new creation, which amounts to one-third of a face. We begin with 2 faces and end with $4\frac{1}{3}$.

In our account of sides, we begin with $2m$ and by division we obtain $(2m + 8)$. Newly created faces are to be credited as 6-gons on average, so our actual total will be $(2m + 10)$, allowing for a one-third share in a 6-gon.

Exactly as in previous operations, we have now to obtain the average number of sides for the faces existing before and after our dividing operation, and equate that average to the general average for the whole tissue. We have:

$$\frac{2m + (2m + 10)}{2 + 4\frac{1}{3}} = \frac{72}{14},$$

$$m = 5.64286 = \frac{72}{14} + \frac{1}{2} \quad \text{exactly.}$$

A face newly formed as a result of division therefore has the average polygonal grade:
$$\tfrac{1}{2}(4 + 5.64286) = 4.82143.$$

This establishes the necessary circulation of face forms which has to go on if a tissue is to remain in steady state and not embark upon any one-way course of histological differentiation.

Face form and face area

We have just obtained decimal limits which, in terms of averages, define the life-course in the tissue of any one cell face, from its formation by division of a pre-existing face to its disappearance in a further act of division. The life-course for faces which arise as new creations is different.

The life-course of a face is relatively short and covers only 0.82143 grades in the scale of polygons, not much more than half the range of $1\frac{1}{2}$ grades which we found for polygons which are sections of cells. The reason for this difference is clear: we have now a steady influx of 6-gons etc. which, in relation to all the averages in the fourth column of Table 2, must be regarded as high-order polygons. Because there is this constant addition of high *n*-gons to the system, the need to upgrade other faces between their successive divisions is reduced.

The effect of face-creation can be seen in another way by using again the concept of loans and repayments employed at p. 42. For the division of a face we have now (in terms of excesses and deficits from the average level of 5.14285):

$$(\tfrac{1}{2} \text{ side borrowed}) \rightarrow (1 \text{ face division}) \rightarrow (\tfrac{9}{28} \text{ side lent}).$$

This is not balanced, and does not need to be, because the newly created faces are sufficiently rich in sides to make up the deficiency.

It is desirable that we should relate our newly discovered decimal limits to the range of polyhedral forms. Reference to Table 2 will show that a polygonal grade of 4.821, which we now take to be the average starting-point for a cell-face in the circulatory system, is nearest to the average for a 10-hedron, though we know from p. 64 that the starting-point for a cell is the $12\frac{1}{2}$-hedron. Is this not a serious discrepancy? Similarly from Table 2 we find that 5.643, which we take to be the end-point for the independent existence of a cell-face in free circulation, must relate to a 35-hedron or thereabouts. As we know from p. 64 that a cell will normally divide at about the 17-hedral level is this not totally destructive of some part of our theoretical development?

In order to resolve this problem we introduce

$$f = \text{specific face area} = \frac{\text{area of any selected face}}{\text{average area of all faces}},$$

exactly in parallel with our definition at p. 64 of the concept of specific cell volume. We now require a practical calibration which will link f with n, the polygonal grade of the face.

A purely algebraic solution is not available, owing to the complexities of a system which is in part unidirectional, in part circulatory. We resort therefore to a method of graphical approximation.

We prepare a graph of f against n, on which three points are to be entered. The middle one is $(n = 5.143, f = 1)$ and can be treated as fixed. The others relate to our decimal limits for the cycle of face division. They must of course refer to specific areas in the ratio 2:1, and we know that if we had only a circulatory system to reckon with then two faces after division and one before division would have to average to unity. That is to say, we would have the two further points: ($n = 4.821, f = 0.75$), ($n = 5.643, f = 1.50$). Because the system contains a process of new face creation these points cannot be quite right. Let us nevertheless enter them provisionally and inspect the result.

When this is done we find (which could easily have been anticipated) that the line through the outer points runs a little too high to take in the centre point. What has happened is that the continuous injection of high-numbered (therefore large) faces very slightly inflates the overall area average. To correct for this we need to reduce a little our $f = 0.75, f = 1.50$. Nothing we can do will get our three points into an exact straight line, but the adjustment needed is so small that if we substitute $f = 0.72, f = 1.44$ we obtain a very smooth curve with little departure from linearity up to about $n = 5\frac{1}{2}$.

For a treatment of the problems of high polyhedra we shall need to extrapolate very slightly beyond this range, to $n = 5\frac{3}{4}$ or thereabouts. In this extrapolation we are guided by the consideration that the graph will have to pass very close to the point ($n = 6, f = 1.65$).

For consider the average area for 2 faces before division and $4\frac{1}{3}$ faces after division, working in exact parallel with our calculation at p. 76, but now using f instead of n. If our calibration curve were perfect the value of this average area would be exactly unity. From the graph-points proposed above we have:

> As to faces,
> 2 before, $4\frac{1}{3}$ after, division.
> As to areas,
>   before division 2 faces at $1.44 = 2.88$
>   after division 4 faces at $0.72\ \ \ = 2.88$
>   and by creation $\frac{1}{3}$ face at $1.65 = 0.55$
>   _____
>
>   Total   6.31
>   _____

Average $6.31 \div 6.33 =$ virtual unity, as required.

It appears, therefore, that we are in command of an imperfect but still quite serviceable calibration. The first use we make of it is to add the final column of Table 2, showing the mean specific areas attributable to the faces of the various polyhedra.

We are now in a position to reassess the significance of the questions raised earlier in this section. If we consider a face just coming to the point of division, it has on average an area of 1.44, which is only 26% above the general level for all the faces of a 17-hedron. If we consider a face just formed by division it has on average an area of 0.72, which is just 20% below the general level for all the faces of cells of polyhedral grade $12\frac{1}{2}$. It seems basically reasonable to suppose that the smallest faces of a cell (in a system where any kind of dynamic equilibrium prevails), will generally be those which have recently arisen from the division of others, and which are themselves not immediately likely to divide again. Conversely the faces of a cell which are most eligible for subdivision would appear to be those with an above-average area for that cell. Our working suggests that an allowance of $\pm 25\%$ in individual face area will largely account for these selective aspects of face behaviour and it is difficult to see any reason for regarding such an allowance as excessive or intrinsically objectionable.

Surface and volume in large polyhedra

At p. 65, and in connection with Table 1, we experienced some difficulty in relation to the specific volumes of cells beyond the grade of 15-hedron or thereabouts. Now that we know more about the areas of cell faces it may be useful to look at this problem again.

For very large values of N we might now reasonably expect to have:

$$V = k \left(\frac{Nf}{14}\right)^{\frac{3}{2}}$$

where k is a constant greater than 1, such that:

$$k = \left(\frac{\text{surface area of average 14-hedral cell}}{\text{surface area of sphere having volume of average cell}}\right)^{\frac{3}{2}}.$$

We have no record of any direct attempt to determine by measurement the ratio between the surface area of a cell and that of its equivalent sphere, and any trigonometrical derivation from a selected polyhedral figure would be open to serious objection. For the time being, therefore, k must be treated as an empirical constant.

In Fig. 30 four observational points derived from the data of Table 1 have been disposed between our linear equation introduced at p. 65 and the new 'spherical' equation calculated on the basis of $k = 1.5$. This produces a superficial appearance of conformity to expectation. We can see clearly how the measurements for highly polyhedral units swing progressively away from one theoretical line towards the other.

In strictly quantitative terms, however, the situation leaves much to be desired. If $k = 1.5$, then the surface area of an average 14-hedral cell must be 31% greater than that of a sphere of the same volume. But even for the deformation of a sphere into a cube the necessary increase in surface is only about 21%, and a cube must be a substantially worse approximation to a sphere than even an irregular 14-hedron. From this point of view the largest reasonably plausible value for k would seem to be about 1.25. Such a low value is not compatible with the existing data on the volumes of

30-hedra, and it must be seriously questioned whether these measurements, derived as they were from shot and bubble experiments with a highly unnatural population structure, are suitable for inclusion in this kind of calculation. If we restrict ourselves to observations up to the 26-hedron, then all the data will fit $k = 1.2$ or even a little less. If $k = 1.2$, then the

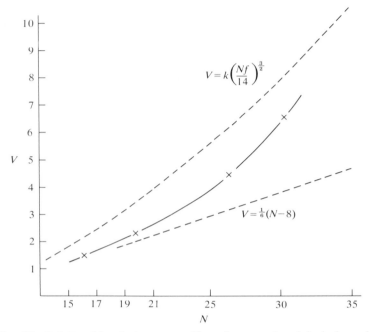

Fig. 30. Relationships between specific volume and polyhedral grade. Four observational points (crosses) fall between a linear equation (lower broken line), derived from the mitotic cycle of shapes, and a 'spherical' equation (upper broken line) which treats high polyhedra as approximations to spheres. To produce this result it has been necessary to give the constant k the value of 1.5, and this calls in question the validity of certain modelling techniques (see text p. 79).

surface of an average 14-hedron is 13% greater than that of the equivalent sphere. If we express this in linear terms, the stretch of the surface metal in a shot-compression model will average at $6\frac{1}{4}\%$. This seems a perfectly credible figure, but any observational work tending very greatly to increase it would certainly require the most rigorous critical examination.

A census of faces

We now require a comprehensive survey of the natural occurrence of faces of the various polyhedral grades. The way in which this survey is to be carried out is in some respects a matter of choice, but to a large extent the procedure is dictated by the forms of publication adopted by the original investigators.

In preparing the tables which follow it has in fact been necessary to take only *one* decision which could seriously affect the scientific issues involved, and this concerns the grouping together of data derived from biologically different sources. It is almost inevitable, from the human and psychological point of view, that each research worker should assiduously strive to make out that his own particular tissue, which he has selected with such care, and examined with such thoroughness, is at least a *little* different from all the other tissues previously reported. The research literature is full of such protestations. All should command a measure of sympathy but relatively few will pass the customary tests of statistical significance. It has therefore been thought justifiable to try the effect of using the observations in ways which their original authors might not entirely approve.

We have data for a large number of plant cells from the interior of tissues which are probably as nearly isotropic as it is possible to find. There is a smaller body of observations for surface cells, which obviously constitute a scientifically distinct problem. In relation to both these situations we can also cite results from compression models. Lastly we have a group of four reports from plant tissues exhibiting a degree of columnar organisation. The separation of this group appears to be statistically justified, but it is unfortunate that the columnar situation has not been effectively modelled. The list of sources is as follows.

Internal units in isotropic systems
 Bubbles: Matzke (1946).
 Cells: Dodd (1944), Duffy (1951), Holtzman (1951), Hulbary (1944, 1948), Marvin (1939*b*), Matzke & Duffy (1955).
 Alloy grains: Williams & Smith (1952).
 Shot: Marvin (1939*a*).

Surface units
 Cells: Matzke (1948, 1949), Mozingo (1951).
 Shot: Marvin (1939*a*).

Columnar tissues
 Cells: Lier (1952), Rahn (1956), Seigerman (1951), Wheeler (1955).

In Table 3 the frequencies have been tabulated to a convention of three decimal places, or number of occurrences per thousand randomly chosen

Table 3. *Frequencies of polygons*

	3	4	5	6	7	8
		Internal units				
Bubbles	000	105	669	221	004	000
Cells	002	276	367	304	047	005
Alloy grains	038	298	367	218	062	000
Shot	057	157	417	277	081	010
		Surface units				
Cells	007	399	422	120	040	010
Shot	035	319	434	175	032	004
		Columnar tissues				
Cells	037	260	349	245	082	023

faces. If it is desired to have numbers on a 'per cell' basis the table entries must be multiplied by 14, or for surface units by about 11.

We postpone discussion of the differences between the various lines of Table 3, and proceed directly to a more complete analysis of the processes which generate cell-face forms in the simplest kind of tissue.

Algebraic models for the generation of cell-face frequencies

The frequencies with which the various n-gons appear as cell faces will depend upon the balance of the relevant events taking place in the tissue. Any specified grade of n-gon may be formed by new creation, by upgrading of lower n-gons, or as a product of division. It may be removed from circulation by upgrading or division. The frequency we observe represents a simple equilibrium between gains and losses. In order to understand the balance of a tissue, we require a set of equations which can be used as a computable model. That is to say, we need to set up an algebraic system which can be used as an experimental subject upon which to investigate the effect of specific alterations in the manner of growth. We cannot control the way in which cells divide, but we can change the coefficients in a set of equations.

We use here a simple form of model consisting of four equations only, relying upon the observational evidence that in the simplest type of tissue only four n-gons need be considered. If p_n is the probability that any randomly chosen cell face will be an n-gon we have very nearly:

$$p_4 + p_5 + p_6 + p_7 = 1.$$

We set out in Table 4 the basic design of our computable model. Initially we know very little of what is to happen: it is the very essence of our task

Table 4. *Derivation of coefficients for a first algebraic model of the events which determine the shapes of cell faces*

	4	5	6	7
4	12	6	—	—
5	6	6	6	—
6	—	12	—	6
7	—	12	12	—
Compensation	$-12p_7$		$+6p_7$	
New creation			1	

For explanation see pp. 82–4.

to make guesses which can afterwards be corrected in the light of experience.

To avoid inconvenience from fractions we enter our table from the left with a notional average sample of twelve cell faces which are to be variously transformed in relation to the mitotic division of cells. This input is to be partitioned according to the frequencies: the number of 5-gons coming into our table is $12p_5$, and so on. We know from p. 75 that for every face which is mitotically divided another is upgraded. The total input to our table of twelve faces is thus to be converted into an output of eighteen faces, six upgraded and twelve from division of six. To this we must add another face by creation, so that twelve faces come into our system of transformations and nineteen go out.

In order to get our model working, we now make three simplifying assumptions:

(*a*) Every newly created face is a 6-gon.
(*b*) Every division of a face is as nearly as possible an equal division.
(*c*) Faces are divided and upgraded in proportion to their frequencies.

On this basis we can at once write three lines of coefficients. In the first line we have 12 because half our input of ($12p_4 \times$ 4-gons) is to be divided equally, and equal division of a 4-gon gives ($2 \times$ 4-gons). We have 6 because the other half of our 4-gon input is to be upgraded to 5-gons. In the second line the distribution is different because a 5-gon cannot be divided equally but only into (4-gon + 5-gon). In principle we would wish to continue in this way and to derive an equation from each column starting with:

$$19p_4 = 12p_4 + 6p_5.$$

In reality we cannot obtain a workable model by simple extension of our existing principles, because we have at present nothing to prevent the upgrading of 7-gons into 8-gons, and if we allow this to happen our model will have a leak in it. Part of the material will escape to high polygonal grades in a completely unnatural manner, and our modelled frequencies will not sum to unity but to something less.

We therefore introduce an arbitrary ruling that 7-gons are to divide but can never be upgraded, so our input of ($12p_7 \times$ 7-gons) is now to give rise to ($12p_7 \times$ 5-gons $+ 12p_7 \times$ 6-gons), thus ensuring $p_8 = 0$ as required by observation.

This decision unbalances our table because the necessary equality between the number of face-dividing events and the number of upgrading events no longer exists. Compensatory adjustment is therefore required. We have introduced $6p_7$ extra divisions, so $12p_7$ division products must be debited to some other column. We have eliminated $6p_7$ events of upgrading, so a counterbalancing credit entry must be made elsewhere. As with other features of our design, the placement of these adjustments must be conjectural at this stage. We proceed on the basis shown in the line 'Compensation' which has been entered in Table 4. Our first attempt at a model is then:

$$19p_4 = 12p_4 + 6p_5 - 12p_7$$
$$19p_5 = 6p_4 + 6p_5 + 12p_6 + 12p_7$$
$$19p_6 = 6p_5 + 18p_7 + 1$$
$$19p_7 = 6p_6$$

The solution to this, with observed frequencies given in brackets for comparison, is:

$$p_4 = 0.221 \ (276)$$
$$p_5 = 0.427 \ (367)$$
$$p_6 = 0.267 \ (304)$$
$$p_7 = 0.084 \ (047)$$

We have succeeded, therefore, in producing four frequencies in the correct order of magnitude. Beyond that, our model has serious defects, as was only to be expected in a first trial. The most obvious fault is the exaggerated prominence of p_5.

The cause of the difficulty lies in the conditions we have set for the division of 6-gon faces. By decreeing that there shall be only equal division we have at a stroke prevented the 6-gon from propagating its own kind and compelled it to breed 5-gons exclusively. We now find this stipulation to be unnatural: except by permitting a proportion of 6-gons to divide unequally into (6-gon + 4-gon) it is hard to see how we can ever hope to maintain in a model the proportion of 6-gons which is found in real tissues.

Here we may note one of the conveniences of an algebraic model, in that it enables us to trace the sources of supply of any particular n-gon. Thus in our present model the 5-gons which are in circulation can be attributed to four sources:

From upgrading of 4-gons	70
From division of 5-gons	135
From division of 6-gons	169
From division of 7-gons	53
Total	427

all per 1000 total faces. The 5-gon population of this model therefore will be very sensitive to any change in the arrangements for division of 6-gons.

We have so far ruled that the division of $(6 \times 6\text{-gons})$ shall yield $(12 \times 5\text{-gons})$. Let us replace this yield by $(2 \times 4\text{-gons} + 8 \times 5\text{-gons} + 2 \times 6\text{-gons})$ and recalculate. Our model is now:

$$19p_4 = 12p_4 + 6p_5 + 2p_6 - 12p_7$$
$$19p_5 = 6p_4 + 6p_5 + 8p_6 + 12p_7$$
$$19p_6 = 6p_5 + 2p_6 + 18p_7 + 1$$
$$19p_7 = 6p_6$$

and gives us:

$$p_4 = 0.248 \ (276)$$
$$p_5 = 0.375 \ (367)$$
$$p_6 = 0.287 \ (304)$$
$$p_7 = 0.091 \ (047)$$

which must be regarded as a substantial improvement. The three largest frequencies have all moved in the required direction and the excess of 5-gons is now a mere 8 faces per thousand.

The main point of concern is now p_7, the value of which in the model is nearly twice that in tissue. Evidently some restraint, not of course amounting to a complete prohibition, must be placed on the upgrading of 6-gons. In relation to the design of a third model we are in fact in a position to lay down a number of guiding principles. The following suggestions seem to present themselves:

(*a*) From an input of $(12 \times 6\text{-gons})$ it is likely that there ought to be roughly three upgradings and nine divisions in place of the present six and six.

(*b*) In introducing extra divisions of 7-gons we made a compensatory debit entry against divisions of 4-gons only. This now begins to look like a mistake, and it might be advisable to check the division of 5-gons a little instead.

(*c*) It was probably right to stipulate that new creation of faces should generate an almost pure flow of 6-gons. To substitute a mixture of 5-gons, 6-gons and 7-gons could only cause a deterioration in our existing model.

(*d*) As regards the division of 4-gons and 5-gons it does not appear that we shall have any reason to impose the kind of inequality which we have been compelled to introduce into the division of 6-gons.

Further improvement of the modelling technique is evidently possible, but will not be pursued here as we have already extracted the principal lessons which can be learned about the parts played by the various *n*-gons. These may be summarised thus:

(*a*) The 5-gon occupies a central position, with several major sources of supply to maintain its abundance, and enjoys a virtual guarantee of numerical supremacy.

(*b*) The naturally occurring frequency of 6-gons is substantially higher than could be the case on a basis of uniform division into (5-gon + 5-gon). Even if all new creations are 6-gons it is still necessary that there should be (6-gon + 4-gon) divisions of pre-existing 6-gons.

(*c*) The formation of 4-gons in substantial numbers is inescapable, and the 4-gon is geometrically unique as the only *n*-gon which is self-propagating in a pure state by equal division.

(*d*) Except by unequal division of a 4-gon or by exaggerated inequality in the division of some higher *n*-gon, there is no way in which the 3-gon can arise at all.

(*e*) Conditions in natural isotropic tissues are strongly adverse to the upgrading of faces beyond the 6-gon level.

The allocation of faces to cells

We have so far considered cell faces as freely circulating mathematical entities. This has not been an unrewarding basis of enquiry, but we have always known that it could be no more than a partial truth. We now have to ask how much can be firmly established about the attachment of particular faces to particular cells, having already recognised at p. 53 that a complete solution is out of our reach for statistical reasons. There are various possible ways of proceeding, and it must be to some extent a matter of opinion as to how far it is justifiable to go. The treatment offered here is built upon two quantities which can be calculated from the data already introduced at p. 81 and which are tabulated in Tables 5 and 6 respectively.

We first define the term *constancy*: the constancy of the *n*-gon is the probability that a randomly chosen cell will have *at least one* face with *n* sides. Table 5 being computed to three places of decimals we have a conventional 999 for universal occurrence (i.e. total constancy). The main result of this exercise is the very curious discovery that in all ordinary

Table 5. *Constancies of polygons*

	3	4	5	6	7
		Internal units			
Bubbles	000	800	999	917	006
Cells	027	999	988	981	436
Alloy grains	352	999	967	857	505
		Surface units			
Cells	059	999	966	705	250
		Columnar tissues			
Cells	427	996	996	990	744

tissues the 4-gon, which for frequency takes only third place among the various *n*-gons, has total constancy, and is the only face-form in that position. This is strikingly displayed in all reported tissue samples outside our columnar group, though strangely overlooked by the original observers. We have for isotropic tissues:

(*a*) 4-gons are 25% less common than 5-gons.
(*b*) About 1 cell in 80 has no 5-gon face at all.
(*c*) No cell without a 4-gon face has ever been seen.

No explanation for this appears to be available.

We may further define the term *gradient*, which is a concept derived from regression analysis. To obtain gradients we first ascertain the numbers of 3-gons, 4-gons etc. per cell for different grades of *N*-hedron, and then calculate the rate of increase for each *n*-gon for unit increase of *N*. For example if the 14-hedral cells in a tissue averaged exactly 5×5-gons apiece and the 17-hedral cells $6\frac{1}{2} \times 5$-gons, this would indicate:

$$\text{gradient of 5-gon} = \left\{ \frac{6\frac{1}{2} - 5}{17 - 14} \right\} = +500 \text{ in 3-figure table.}$$

In the necessary calculations the available data are spread more thinly than in the previous work and the gradients in Table 6 must be used with caution. It appears however that three conclusions are inescapable, though again they have strangely escaped attention hitherto. We have:

(*a*) Internal cells have a higher gradient for 6-gons than for any other *n*-gon.

(*b*) In isotropic tissues the gradient for 4-gons is very low, relative to the frequency of 4-gons. The outfit of 4-gons possessed by a cell must therefore change little during the mitotic cycle.

The faces of cells

Table 6. *Gradients of polygons*

	3	4	5	6	7
		Internal units			
Bubbles	000	+024	+003	+923	+051
Cells	+012	+037	+218	+464	+242
Alloy grains	+013	+037	+301	+396	+196
		Surface units			
Cells	+036	−126	+521	+306	+178
		Columnar tissues			
Cells	+009	+101	+252	+326	+209

(*c*) The gradient pattern in surface cells is grossly disturbed (this is attested by close agreement of all the reported samples).

These would appear to be biological discoveries of some importance, but they are at present purely empirical. It is difficult to see any basis from which any prediction of these results might have been ventured. Nor is the task of retrospective explanation a simple one. It is easy to suggest that (*b*) above should relate to the high constancy of 4-gons and that (*a*) is no more than a consequence of the exercises in polyhedral construction at p. 52, but when an attempt is made to reduce these ideas to a strict logic they prove unexpectedly elusive.

The significance of extreme polygons

It will be obvious from Table 3 that the differences between the various classes of material displayed there are expressed most plainly in the frequencies of the more marginal polygons, specifically the 3-gon and the 7-gon. In respect of these two figures the first four rows present a consistently graded sequence. With shot at one extreme, bubbles at the other, and a large assortment of natural tissues coming between, it is almost impossible to avoid the speculation that abundance of 3-gons and 7-gons may relate to some physical quality of the general nature of stiffness, viscosity, or frictional resistance.

To give such a hypothesis any more definite form will not be an easy task, and it would seem to be urgently desirable that the practice of compression modelling should be extended into new areas of enquiry (the compression of lubricated shot is plainly indicated, for example). The state of affairs in columnar tissues obviously ought to be modelled both algebraically and in shot.

Setting aside the 2-gon (everywhere an extreme rarity) we may note the unique theoretical significance of the 3-gon. Of all the *n*-gons which circulate in tissues, the 3-gon is the only one which relies for its very existence exclusively upon very unequal division of faces. The most symmetrical event which can give rise to a 3-gon is the division of a 4-gon into (3-gon + 5-gon). The 3-gon also has a special relationship with the problems of intrusive growth discussed at p. 55. We noted at that point that there was strictly no law of conservation for cell faces because it was not possible formally to exclude the development of supernumerary faces by intrusive action on the part of some of the cells. The theoretical difficulty is perhaps no longer a source of anxiety, but it is worth observing that a principle of conservation *does* operate in a tissue where the 4-gon is the lowest permitted grade of face. Kelvin (1887), in connection with the development of his minimal 14-hedron, exhibited a soap-film model in which a 4-gon face can be caused by a puff of air to turn round by a flip-flop action. This model is the exact logical counterpart of Fig. 16*a* & *b*, and it embodies a law of conservation which can be evaded only by the introduction of a lower grade of face.

6

The geometry of differentiation

The steady-state concept from which so many of our equations have been derived is simply a mathematical model of a tissue for which it would be correct to write:

Level of histological differentiation attained $= 0$

so it is not difficult to see the desirability of extending our ideas, if possible, a little beyond the stage reached in previous chapters.

There would be little merit or satisfaction in finding *ad hoc* solutions to particular cases and instances. We know beforehand that the more complex tissues display an immense diversity of differentiating processes, and for every intricate relationship which is known to us at the present time we have to expect that many others will be discovered in the future. Our objective therefore ought to be the establishment of some generalised mathematical framework which would be capable of universal application, acting as a standard histological calculus available at all times for the treatment of any new problem which may present itself. The development of such a system is a considerable intellectual enterprise. On the other hand it is fairly easy to see how one ought to begin, and there are some aspects of the matter which can already be reduced within the compass of a student exercise.

We shall begin by proving that any measured quantity which expresses the degree of differentiation of a tissue must belong to one of four mutually exclusive geometrical categories. These categories may be called:

(*a*) Anisotropy
(*b*) Polarity
(*c*) Wave propagation
(*d*) Idioblastic classification of cells.

No matter how complex may be the departure of a tissue from the steady-state model, every mathematically distinct item of that departure must belong to one of the listed categories, and to no other.

Consider the situation of an infinitesimally small observer who is put into the tissue with instructions to maintain a specified velocity (i.e. a fixed speed in a fixed direction) and to supply regular estimates of some specified

90

quantity, for the measurement of which he carries an appropriate instrument. The flow of estimates received from such an observer will always be subject to variation of a statistical nature, arising from the finite size of his successive samples. We may elect to have very frequent but inconsistent reports or more consistent but less frequent ones. In a steady-state tissue there is no other source of variation, or of difference between the reports of one observer and those of another. Unless the speed is high enough to produce relativistic effects, observer velocity is immaterial. In the steady state we may suddenly instruct an observer to double his speed, or to retrace his path in the opposite direction, or to make a 90° change of course, or simply to stop where he is, and his subsequent reported estimates will show no systematic change whatever. We shall simply have no way of telling from his results whether he has obeyed our movement orders or not. In other words we might have an army of such microscopic observers traversing the tissue in all directions, but in the steady state all their reports would constitute a single statistically homogeneous flow of information.

As soon as the reports of our observer begin to display variation of some kind which is not attributable to sampling error, then we shall know that the steady state is at an end. From this point of view the differentiation of a tissue appears simply as a progressive diversification of the measurements which can be taken from it. In particular we may note that the degree of complexity which a tissue has attained is indicated merely by the number of *truly independent* numerical quantities which can be derived from it. Consider for example the question of average cell diameter. In a steady-state tissue this is in principle a single unchanging number. We may of course adopt various measuring techniques. Diameter may be measured between vertices, between the mid-points of edges, or between the mid-points of faces, and there is endless scope for discussion about the relative merits of different systems of computation for the derivation of an 'average'. But in the steady state all diameter-measuring procedures will be interchangeable by the application of conversion-factors. Diameter between faces and diameter between vertices cannot be independent quantities but must stand in a fixed ratio, and it is not necessary to our present argument that we should be able to say what that ratio is.

But now suppose that we find a systematic relationship between the direction of an observer's movement and the numerical estimates of cell diameter which he transmits to us. The tissue has become anisotropic. The cells are elongated consistently along a selected geometrical axis. Cell diameter is no longer a single fundamental quantity or dimension of the tissue. Numerical expression of cell diameter is no longer complicated only by sampling errors or by technical questions about the choice of a scale of measurement, but has come to contain a new element of information. Furthermore this new item of information is purely biological: the direction and extent of elongation are incapable of being predicted from abstract mathematical principles and can only be the result of decision-making processes in the organism. Again let us suppose that an observer on a given line of advance reports that he is crossing cell-faces more or less perpendicularly, while another observer on a different course is experiencing more oblique forms of transit. We now have a different manifestation of anisotropy; the cells have assumed a somewhat prismatic form. No longer do the interfaces in the tissue constitute a single population. Some of them have acquired a special status as the 'ends' of cells, and a degree of columnar organisation must therefore exist. Once more we have proof of a decision-making process in the material.

Anisotropy is a very simple concept from the purely mathematical point of view, and it is most improbable that we would ever have much trouble in devising suitable forms of measurement. For most purposes a length:breadth ratio, or something equally straightforward, is all that is ever likely to be needed. For more detailed theoretical interpretation, if that should be required, it is obvious that one would have to turn in the first instance to the ellipsoid as the parent figure (consider, for example, the analogy with wave-fronts in the propagation of light through a birefringent crystal). It is easy to imagine ellipsoids prepared in clay or similar material and used in a compression model: the 'cells' could be given any desired ratio of polar to equatorial diameter and arranged with consistent orientation at the beginning of the experiment. There seems to be no public record of any such trial.

So far as our imaginary observers are concerned, a polarity phenomenon is simply one in which the nature of the observations is changed when the line of advance through the tissues is exactly reversed. An example would be a consistent voltage-drop across cellular interfaces. If one observer on a straight-line course finds each new cell to be at a lower voltage than the preceding, then his colleague coming in the opposite direction will have a contrary experience. Polarity is perhaps the easiest to understand of all the possible departures from the steady state. The mere recording of an isolated manifestation of polarity is not a matter which can ever call for special care or resourcefulness from the mathematical point of view (it may, of course, present great practical difficulties in the laboratory, but that is a separate issue). It has, however, to be noted that polarities are unique in their manner of combination. Wherever we have three polarities on independent axes there will arise one of those situations which must be characterised as right- or left-handed and which is not geometrically congruent with its own reflection in a plane mirror. There is no other combination of circumstances which can make it necessary to distinguish between stereoisomerous configurations (Fig. 31).

So far as the nature of observations may be consistently related to particular directions of travel through the tissue, we have therefore only these two rather simple concepts of anisotropy and polarity. Neither of them can present any particular difficulty of mathematical formulation, and neither will require extended discussion in a book of this nature. We shall be able to treat them mainly as basic ingredients in various more complex patterns of development.

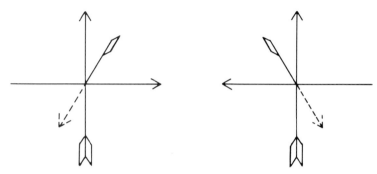

Fig. 31. Three qualitatively distinct polarities are represented by arrows differing in their state of feather. Two lie in the plane of the page while the third strikes downward through the paper. When polarities are assembled in this way the mirror-image figures are non-interchangeable, no matter how they are turned about in space.

The differentiating processes which are of more serious geometrical interest are those in which there begins to be statistical heterogeneity even among the observations associated with a single direction of travel through the tissue. One of our observers, let us say, moving along a straight track, sends us reports of cell diameter. We plot these as a frequency distribution, and the curve turns out to be bimodal. We have to find the general principles governing the interpretation of this class of data.

Evidently the tissue is segregating into distinct types of cell. Whereas in the steady state the difference between one cell and another is merely a transitory accident of mitotic phase, we now see the emergence of morphological distinctions of a more substantial and lasting character. Whereas polarity and anisotropy are only differences between geometrical axes, and can therefore exist in structurally uniform tissues, we now have to deal with phenomena which will always imply the existence of a morphological classification of cell-types. Furthermore, we shall have occasion to prove (pp. 115–23) that any such classification is bound to be extensive: as soon as a tissue contains more than one cell-type it cannot stop at any small number of types, but automatically jumps to an altogether higher order of complexity.

We have therefore to envisage a situation in which, among a mass of initially indistinguishable cells, certain individuals become 'labelled' as members of a particular type, acquiring structural and biochemical characteristics which set them clearly apart from their neighbours. It is not necessary that such a 'label' should be absolutely permanent, and indeed we shall get on much better by treating many aspects of differentiation as potentially reversible even where the organism does not live long enough to experience the reversal. We need stipulate only that the segregation of a tissue into cell-types should be a more persistent effect than any incident of the mitotic cycle.

It is biologically inconceivable that the member cells of any morphologically distinct type should be randomly distributed in the tissue. Except by endowing the interfaces with a lethal degree of impermeability there is no way of excluding such effects as the localised depletion of supplies, the circulation of hormones, or the outward spread of waste products. In practice, therefore, one of two things must always happen. Either a cell which adopts a specialised course of development will tend to be closely associated with neighbours of its own kind, or alternatively there will be statistical evidence of repulsion, amounting perhaps to a rule that no two cells of the specified type can ever be adjacent. We are not at all concerned with the physical causes of these phenomena, but only with the contrast between two statistical distributions, of contagious and repulsive character respectively. Between these two lies a neutral state which is of no interest to us because no biological example of it can be found.

Where the statistical situation is one of contagion or of the grouping together of similar cells, then the only suitable forms of mathematical expression are those of wave-mechanics. If cells are seen to behave in a certain way, and if that observation increases the probability that similar behaviour should appear also in the next cells, then what we are witnessing is purely and simply the propagation of a wave. It makes no difference whether the process is determined by repeated individual stimulation of one cell from another, or by a general influx of hormone from some external source. Either way, a wave is what we shall actually *see*. Wherever the differentiation of a tissue is governed by wave equations we must expect to see contrasting cell-types arranged in some kind of patchy or banded pattern. Developmental studies (see p. 98), when pursued with sufficient energy, will often show that these patterns are travelling through the tissue so that they must be treated as crests and troughs directly comparable with those of surface waves on the sea. In other cases the patterns are essentially static, and must be regarded as standing waves or interference fringes.

Many biologists are extraordinarily hesitant about the application of wave concepts to the growth of organisms, although the fund of elementary examples is really quite inexhaustible. Every striped or spotted animal is a mobile demonstration of standing waves closely comparable with those which physics teachers show to their pupils by the use of ripple-tanks or acoustically vibrating metal plates, and it is easy to find such little refinements as the difference in wavelength which distinguishes between certain zebra species. Again it should have been obvious, but biological writers mostly have not found it so, that the growth of an organism is likely to generate rectified wave-forms. Apply a sinusoidal voltage to a selenium rectifier and you get intermittent d.c. Similarly apply a sinusoidal temperature regime to a tree and you get intermittent growth as displayed by the annual rings (and with innumerable analogies in fish scales etc.). A tree or a fish is not quite like a battery charger, but on the other hand the normal rules of Fourier analysis are not going to be suspended just because we have to deal with a living subject. We must therefore expect a substantial body of histological theory to be more or less common ground with such disciplines as electronic engineering.

When, on the other hand, the members of a particular cell-type exhibit statistical evidence of a mutually repulsive tendency, wave propagation becomes impossible. Where there is no concerted action among the cells there is no way of building up anything which could be treated as a crest or trough. Morphological differences are not regional but individual. We then have a situation in which the cell is sharply and permanently distinguished from its immediate neighbours so that it will qualify for recognition as an *idioblast* (see Chapter 7). From the analytical standpoint the significant distinction is that relating to the distance over which influences are likely to be propagated. When the example of one cell is followed by its neighbours it is possible and indeed commonplace for the resulting system of mutual reinforcement to acquire an energy package too great to be dissipated within a confined space. In the colouration of the zebra and in some of the processes occurring in treetrunks we have wave-fronts sufficiently energetic to traverse large organisms virtually from end to end, transforming millions of cells on the way, yet suffering no obvious weakening or attenuation. Where there is no reinforcement the influence of one cell, which cannot be very great to begin with, must soon be diluted or attenuated by hostile or indifferent neighbours. The volume of tissue within which it is profitable to seek observational evidence of interaction is then very restricted, a radius in all probability of no more than three or four cell-diameters away from the original disturbance. A small energy-package, radiating into an uncooperative or resistant surrounding medium, cannot be expected to produce conspicuous effects at great distances.

There is here a beautiful and fruitful analogy with the great distinction which exists in theoretical physics between particle physics on the one hand and mass-action physics on the other. It is a question, primarily, of the absolute scale of magnitude of the phenomena to be examined. In so far as we may choose to study wave-regulated aspects of differentiation it may be more of a hindrance than a help to look closely at individual cells. Much that concerns differentiation is distinctively of the nature of statistical generalisation, to be compared with such physical laws as those of Boyle and Ohm. For aspects of differentiation in respect of which cells do not interact cooperatively we are however compelled to follow the physicist in abandoning wave concepts for particulate ones. When cells of a given type are distributed upon a principle of mutual repulsion our attention is concentrated upon a relatively tiny sphere of action, and the laboratory record assumes the form of a double catalogue. We have firstly to distinguish all the cell-types present, and secondly to ascertain the consequences of each of the possible kinds of cell-to-cell encounter.

It will be found logically impossible to conceive of any form of differentiation which is not embraced in the foregoing considerations. We have provided for every occurrence of heterogeneity in the reports in our little observers, and so established the primary classification of phenomena introduced at p. 90. We have naturally to expect that any real tissue will simultaneously display all four categories of effect. What we have done is to distinguish between mathematical methods appropriate to different purposes, and there is no inconsistency in planning to use wave concepts for one set of measurements and particle concepts for another, even though the same specimen is employed throughout.

Before we turn to the examination of actual tissues it must be noticed that although our four categories of differentiation are mathematically separable they are not independent, but interconnected in various ways which will be more or less self-evident from familiar physical analogies. A single cell which adopts a distinctive course of development automatically polarises all its neighbours in a concentric pattern, just as a magnetic pole imposes its influence on a surrounding population of iron filings. A hormone entering an anisotropic tissue will travel with different velocities in different directions, and the wave-front therefore assumes a form for which the appropriate mathematical treatment is already extant in text-books of crystal optics. Again it must be obvious that any physiological influence having the mathematical structure of a wave will polarise the cells through which it passes, and also that the intensity of that polarisation will depend on the ratio of wavelength to cell diameter. To appreciate this, consider a ship heading into the waves: the fore-and-aft pitching moment is maximal when water-line length is half the wavelength so that the bows are on a crest as the stern drops into a trough and *vice versa*, whereas longer or shorter waves have less effect. Simulate the maturation of a cell by freezing the sea, and one has a geometrical relationship which is immediately relevant to many histological examples. For principles of such elemental simplicity as these the reader will probably be able to draw suitable biological illustrations from his personal experience of actual tissues. Convenient examples can be found in the surface tissue of roots, where a wave of progressive vacuolation can be seen to enter an elongated cell from its basal end. Such cells commonly mature before the wave has passed right through them, and the polarised distribution of cytoplasm exerts a direct and obvious influence upon the placing of hairs and mitotic septa.

Whatever may be the exact line of thought that an individual student may choose to follow, the mathematical treatment of differentiation will always resolve itself into two sharply contrasting divisions. In so far as we may have concerted action by many cells, a mass transformation of tissue, then we stand upon the homely ground of wave-mechanics, and we can be quite confident of finding such conveniences as ready-made computer software. A wave in histology is not a *mathematically* distinct entity from a wave in any other science, and the response of a tissue to such influences as an influx of hormone therefore cannot conceivably present us with any fundamentally new form of analytical problem. Of course the practical difficulties of our investigation may be very great, but so far as scientific principle is concerned, our task cannot amount to much more than straightforward algebraic substitution: we shall be writing new variables and constants into equations which have had a long history of previous application in tidal forecasting and the design of electrical machinery, seismology, meteorology, the tuning of bells, and various other fields of activity. Tissue differentiation involving mass action is therefore not a study in which we can hope to discover anything which will be unfamiliar or novel in any mathematical respect. Biologically such enquiries may be infinitely rewarding, but they require only such patterns of thought as are already well established in the mind of every working scientist.

When we come to look at more localised aspects of differentiation and to concentrate our attention upon unitary interactions between one cell and another, a radically different situation arises. In this part of our work it will be necessary to study some geometrical relationships which may fairly be described as obscure, and which seem unlikely to gain any general currency, or to find much practical application in sciences other than histology. For a proper understanding of developmental phenomena both levels of mathematical appreciation are inescapably necessary, and there would be no answer to the suggestion that the cellular level is the fundamental one, and should logically be studied first. The historical development of science, however, does not usually proceed from what is fundamental, but from what is too obvious to be ignored. So it has been in this instance, and for purposes of exposition there may be some advantage in retracing the historical pathway from that which is self-evident to that which is slightly less so.

Examples of wave propagation

Wave effects in histology are of almost universal distribution: the fundamental concepts of wave-mechanics offer a ready-made statistical 'fit' to any tissue mass which is at all regularly segmented, articulated, stratified or corrugated, and can be extended with no particular difficulty to cover a range of more complex phenomena. Layering of a tissue in relation to a boundary (whether an external surface or the lining of a cavity), or the development of patchy or reticulate patterns, or any progressive change in a tissue with the passage of time, can always be reduced to wave equations if we find such a reduction to be convenient or helpful. How far it may be expedient to move towards the formal embodiment of a wave-type algebra will always depend on the number and quality of the relevant observations. Where our knowledge goes no deeper than superficial recognition of wave effects, there is neither the opportunity, nor any immediate need, for exact mathematical expression. That the fin-rays of a fish and the lateral veins of a clover leaflet owe their regular spacing to wave action of one sort or another is clear enough, but further theoretical analysis of these cases seems at present to be out of reach. There has not been the kind of thorough and painstaking laboratory enquiry which would enable us, for instance, to extract observational estimates of velocity or frequency, or to draw a representative wave profile.

The laboratory difficulties of such enquiries must always be considerable, and naturally increase as the effect to be investigated becomes more transitory, more localised, and more inaccessible. Duration is in fact the dominant consideration from the technical point of view. Serious analytical work is only practicable with a wave system which can be kept going for long periods of observation, and it is no accident that the histological waves for which more detailed assessment can be given are all associated with the sustained apical growth of plants. In the development of a stem or root there is a permanent wave pattern which is mathematically analogous to the appearances seen in aerial photographs of ships crossing the ocean, or in the familiar images of shock waves surrounding bullets in flight. That is to say, the stem or root apex in its passage through space is accompanied by an elaborate but relatively unchanging pattern of disturbance. The resulting circumstances are uniquely favourable to laboratory enquiry, and the structure of the mature tissues can be treated, to a quite significant extent, as a kind of solidified interference-pattern.

In one respect only our ship and bullet models are a little misleading, and require qualification. Muzzle velocity of most firearms is about comparable with the velocity of sound in air, so no air-wave caused by a bullet can very much outrun its parent projectile, and similar considerations normally apply to the wake of a ship. By contrast the apical growth of plants is slow, and many histologically significant waves are capable, so to speak, of coming up from behind, running out through the 'nose' of the system, and vanishing into the empty space ahead.

It must be noticed also that in order to affect the geometrical organisation of a tissue, a biological wave-form must have a frequency which is reasonably well tuned to the duration of some formative period in the life of the cells. We have for example some rather controversial evidence that

Fig. 32. Natural wave-form in timber. Specimen is an axe-chip, showing ripples in low relief on a cleft surface under oblique illumination.

the growth of many plant organs proceeds in pulses of a few minutes' duration. If this phenomenon is genuine it is a rectified wave-form, but there can be no reasonable expectation of any geometrical consequence because the reported frequency is too high: any cell in its development will experience so many alternations that the cumulative result will be a statistically balanced average in which the distinction between crest and trough is quite lost. There is no need for timing to be very precise, but it must not be hopelessly inappropriate: one cannot change the going of a pendulum clock by playing a flute in the room, but it can be seriously disturbed by vibrations of lower frequency, even though these bear no regular relationship to the natural swing of the pendulum. Probably the highest frequency which is likely to influence the gross geometrical organisation of a tissue will have a periodicity best measured in hours, and the conditions attaching to the practical examination of specimens will generally make it more convenient to deal with frequencies in a range substantially lower than this.

By far the largest coherent body of observations available to us concerns the cambium in the trunks of trees, and more particularly of coniferous trees. We have here a tissue of fairly simple construction, in which a high

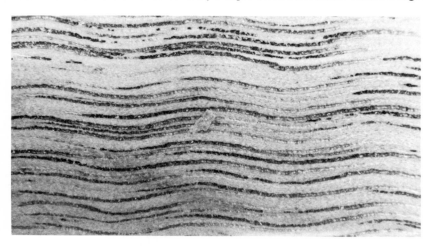

Fig. 33. Natural wave-form in timber. The planed edge of a board is laid horizontally across the page. The grain ripples here are about a centimetre from crest to crest, and they advanced through the living tissue at a rate of about a wave-length every eight or ten years. By taking successive shavings off this surface it would be possible to observe earlier conditions of the tissue. If successive shavings were used as frames in a cine-film, the waves would be seen to move across the screen like water-ripples observed through the side of a glass tank.

rate of mitosis can be maintained for several centuries, and having moreover the peculiar advantage that changes in the cambium are permanently chronicled in the secondary xylem which it produces. The historical development of any cellular configuration which we find in the cambium can be retraced by the simple expedient of cutting serial sections deeper and deeper into the wood, much as an archaeologist may excavate progressively to deeper levels of his site. Some of the wave-forms occurring in this system of tissues are extremely conspicuous, have been known to woodworkers from time immemorial, and have given a natural stimulus to research (Figs. 32–35). Inevitably the investigators concerned have tended to concentrate upon the special features of their own material and much of the literature displays a tacit presumption that the mere occurrence

Fig. 34. Advance of a wave-train through the cambium of a tree. This is a planed surface equivalent to the upper face of the board sectioned in Fig. 33. Apart from a little roughness left by the tool the appearance of relief is an illusion: there is merely a difference in reflection of light according to the slope of the fibres. The wave-train was moving downwards through the cambium on the right-hand edge: as a record of this the grain-ripples are inclined relative to the top edge, which is cut as nearly as possible transverse and serves as a graphical axis of time.

Fig. 35. Natural oscillation in timber grain. A scrap of mahogany rail was split upwards from the bottom by a straight-edged chisel inserted along a radius of the tree. If the split had run true it would have emerged as a straight line joining the white markers. The wave revealed here is similar in principle to those in Figs. 32–34, and of roughly comparable frequency, but its wavelength and velocity are very much greater.

of wave phenomena is in itself in some way remarkable, and to be regarded as a distinctive characteristic of cambial tissue. There is no rational basis for such sentiments. No doubt the principles could be exemplified equally well from animal tissues. Really it is only a question of supply: we have many observations on the structure of deal planks, but very few

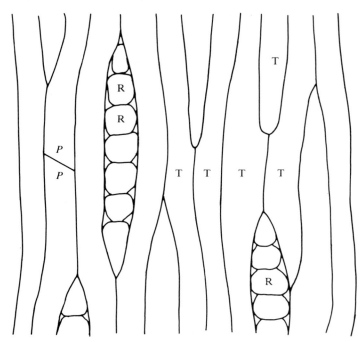

Fig. 36. Longitudinal section of a coniferous timber, consisting of tracheids (T) and ray cells (R). Such a section preserves for us a record of cell outlines in the cambium at some past date. At *PP* two tracheids are very untypically separated by an almost transverse wall. Such partitions do not long maintain this posture but are quickly turned into a much more oblique position, so the configuration is evidence of a very recent 'pseudotransverse' cambial mitosis. Although pseudotransverse division contributes to the increase of cambial area as the tree enlarges, most of this activity merely compensates for a high level of cell mortality.

measurements relating to the formation of stripes in the skin of the foetal tiger. The botanical example therefore takes precedence for the time being.

   The woody tissue of a conifer consists almost wholly of two forms of cell, the elongated tracheids and the much shorter cells of the rays (Fig. 36). The rays are of rather specialised botanical interest and for our immediate purpose we can disregard them and concentrate our attention on the tracheids, the shape of which serves as an automatic record of changes in

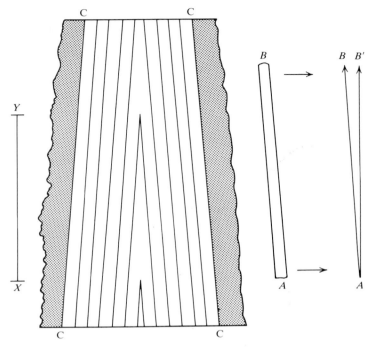

Fig. 37. Relative motions of a cambial wave. At left is a longitudinal section of a tapering tree-trunk. The (shaded) bark lies outside the cambial layer C. The height growth of the tree for one year is shown by XY, its radial growth in the same period is shown by the spacing of the lines in its woody core. A single cambial cell AB stands in an inclined position but is riding outwards on the growing face of the wood. An observer riding with the cell sees a wave enter the cell at A and travel along the axis of the cell to be discharged at B. For an external observer, B is not a fixed point, but will have moved to some such position as B' while the wave-front is contained in the cell. If wave velocity is equal to height growth (of course for the external observer, not the moving one!), AB' is vertical.

the cambial cells from which the tracheids are derived. Location of the cambium in the tree is shown in Fig. 37. It is a thin conical sheet, inside which the wood is organised in concentric smaller conical layers each normally representing the output of one year's cambial mitosis. Because the wood is incompressible, the conical profile of the cambium moves upwards roughly in time with the height growth of the tree. An individual cell therefore moves horizontally outwards as shown in the figure, not quite parallel to its own length. Any wave-front travelling with the shoot apex is consequently subject to the relativity shown in the vector triangle ABB'. The cambial cell experiences such a wave-front as an influence strictly parallel to the tapered outer surface of the wood, whereas an external observer sees the same wave moving in a strictly axial direction.

In these conditions it is a perfectly safe theoretical prediction that the cambium must be traversed by waves of at least two main classes: waves following height growth, with a velocity best expressed in cm per annum, and waves of annual frequency, with velocities which seem likely to fall in the kilometric range. These higher velocities are in fact too great to be measured by conventional histological methods: for annual changes in the cambium there is no particular difficulty in plotting representative wave-forms, but there is commonly near-simultaneity between events at top and bottom of the tree. The very direction of propagation is then ambiguous: one can observe at one time no more than about 50 metres of a wave which is quite possibly a hundred times that length. Waves which accompany height growth, and indeed some which move a good deal faster, are however perfectly amenable to velocity measurement in the laboratory.

As cambial cells are elongated, their degree of elongation is necessarily responsive to wave action. Average length of cell is the variable which is customarily measured. It represents a dynamic equilibrium between a substantial number of conflicting histological changes, but for our present purpose it will be permissible to simplify a little. The main reason for cambial cells to shorten is 'pseudotransverse' division by a sloping wall somewhere along their length (Fig. 36). The main reasons for increase in length are firstly a high rate of cell mortality with enlargement of the survivors to fill the vacant spaces, and secondly the rearrangement of cells to provide for the necessary annual increase in cambial girth. The botanical literature contains not only numerous measurements of cell length but also some observations on the causally related ultrastructure of the cell wall: the more elongated the cell the more the cellulose microfibrils are turned away from a transverse orientation towards a longitudinal one. Assembling the available evidence, it seems that three principal waves are required to regulate the average length of cell.

It has been known for a century that the length of cell increases progressively during the life of the tree. We have a choice of sources for a generalised wave-form; it may be interesting to use the data of Bailey & Shepard (1915) because in fact these authors denied that their observations showed any regularity of increase. But this was only because they attempted a subdivision of time rather finer than the quality of their data would justify, and adhered to absolute values so as to deprive themselves of facilities for pooling observations from different specimens. Simply by combining their measurements on the basis that cell length in the 10th year of life is to be taken as 100, we obtain Table 7, which is a reasonably smooth wave-form. If we take it that a tree begins its life at a minimum, then the wavelength here is of the order of 350 years of height growth, and will

Table 7. *Average length of conifer tracheids*

Calculated from data of Bailey & Shepard (1915) by running four species together on the basis that tracheid length in the 10th year is to be taken as 100

Years	Length	Years	Length
10	100	95	191
20	126	115	194
35	166	150	203
55	169	200	179
75	174		

be more if that assumption is unsound. Waves of this type appear to be universal in conifers: probably no tree ever survives to experience a complete cycle of oscillation. As regards the relative motions shown in Fig. 37 it will be noticed that if the wave exactly keeps pace with the other effects of growth then mean cell length at any given radius outwards from the central axis will be independent of height above ground. To express this a little differently we would expect the same observation in the thirtieth ring of wood from the centre at ground level as in the thirtieth ring high up in the trunk, but of course these rings do not correspond, and the upper observation is not in wood which directly continues the grain of the lower sample. We have a circular wave-front of fixed diameter moving along a system of conical layers sloping inwards: the corresponding relationship for ultrastructural obliquity of the wall has also been recorded (Preston, 1974, p. 320).

Superimposed upon this long slow wave of cell length we have a high-velocity annual fluctuation. It is not difficult to see that wave-forms synchronised with the seasons will show rectification effects: that is to say, crest profile and trough profile will not coincide when one of them is inverted, because winter does not simply reverse the conditions of summer, but has its own peculiar quality. To see a rectified wave in a tree, see Fig. 38 which is a transverse section of coniferous timber with a graph of radial cell diameter. This wave is ratchet-toothed, and although the synchronous wave which affects the longitudinal diameter of these cells is much less perfectly known, it certainly possesses an unsymmetrical quality which is somewhat less extreme though not different in principle. There is in fact a partial separation in time between death of cells with enlargement of others, these effects taking place early in the growing season, and pseudotransverse division, which reaches its peak some weeks later. We could of course express these components of the situation at choice as

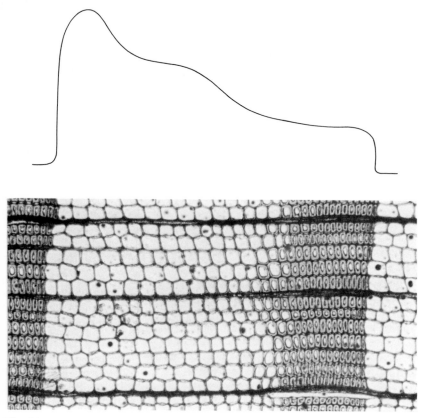

Fig. 38. Wave-form in the radial growth of a tree. Photograph shows transverse section of coniferous timber, growth proceeding from left to right in the direction marked for us by the rays of the stem, three of which appear as dark horizontal lines. Tracheids formed early in the year undergo great radial enlargement, while those formed later do not. Measurement from the photograph yields the curve above, a rectified and very unsymmetrical wave. The graph may appear to conflict with the visual impression of the specimen because it has been plotted by mitotic cycles; the late-season tracheids occupy a large part of the time-base without contributing in the same proportion to the thickness of the tree.

separate waves of mortality and mitosis with appropriate phase-difference. Concentrating, however, upon the average length of cell, what is observed is generally a rather smooth build-up of cell length in the spring, compensated by a distinctly more abrupt autumnal decrease brought on by an epidemic of pseudotransverse division.

If this were all, then the whole history of cell-length in the wood could be expressed by adding a yearly jag-tooth of unvarying form to the longer wave of Table 7. Fortunately there are observations good enough to reveal a more interesting situation (Bannan, 1966). There is in fact, during the life of the tree, a progressive intensification of the annual fluctuation. Both the rate of cell mortality and the late-season concentration of pseudo-transverse division are accentuated with age. This constitutes, in the

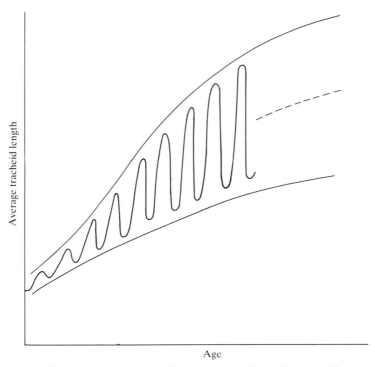

Fig. 39. Draft scheme for variation of average tracheid length in a coniferous timber. There are three component waves: an upward low-frequency trend (broken line), an unsymmetrical annual oscillation with a winter minimum and a late-summer maximum, and an amplitude modulation indicated by the upper and lower curves. This appears to be the simplest pattern which will even approximately fit the available observations.

language of wave-mechanics, an example of amplitude modulation, in which the annual jag-tooth acts as a carrier wave for another much lower frequency, in the manner of Fig. 39.

It would be a simple matter to extract from the botanical literature a considerable array of examples in which the data can be reduced to wave-forms of one kind or another, but there is no point in multiplying

mathematically equivalent biological illustrations in a book of this character. There is nothing distinctively botanical in our theoretical concepts
and even the natural preservation of data which so conveniently takes place
in timber specimens will be observed also in antelope horns, mollusc shells
and the like. Nor do patchy or reticulate patterns involve any idea which
is basically more difficult than the realisation that wave systems may cross
one another at an angle instead of running to and fro along the same
straight line.

We can however advantageously pause to take stock of our position,
and to appreciate that our use of wave-forms so far has amounted basically
to a biological application of Fourier principles. We have been cataloguing
frequencies as a means of ascertaining how many sources are contributing
to a given effect, and we have noted that any unsymmetrical crest-form
necessarily implies a phase-difference between two component waves (e.g.
cell-mortality peaking at a different season of the year from mitotic rate).
All this is very satisfactory as far as it goes, but it leaves some important
questions unanswered. In particular we still lack any concepts which would
enable us to predict the speed of travel of any histologically significant
wave. In fact we have merely been observing the passage of waves, and
the patterns of their interaction, without looking closely at the actual
mechanism of their advance. To examine the fundamental histological
basis of wave transmission is a considerable further step in our analysis,
but even this problem is not absolutely unapproachable. Our attack upon
it is founded upon the generality of wave concepts, and our consequent
freedom to draw upon models from the physical sciences.

Wave dynamics in tissues

We start from the observation that histologically relevant waves are all
slow-moving by physical standards (i.e. in comparison with light, sound,
and surface waves on water). Our primary need therefore is to visualise
the means by which a tissue can retard the passage of waves of differentiation. This can only be a matter of energy relations, because there is no
substance in a wave at all: it is merely a travelling set of energy exchanges.
We have an unlimited choice of models, but a simple electrical system will
serve as well as any.

Consider the traditional type of overhead telephone land-line. This is
as nearly as possible a purely resistive circuit, thin wires widely separated
in air, and regularly transposed on the cross-bars so as to secure a
non-inductive winding. Let us feed a battery voltage into the line through
a rotary commutator giving regular reversals. Any roving observer with
appropriate equipment will be able to detect this applied voltage even at
remote points down the line. We have established a square-wave voltage

travelling at very nearly the speed of light, and arriving at any distant station with its square wave-form almost unchanged. Now suppose that we connect a suitable capacitor between the line-wires at every post. What happens at the next voltage reversal? Battery current will flow down the line, but at the first post part of the current is diverted to charge the capacitor there. A weaker current continues into the next span of wire, but then there is another capacitor to charge, and so on all the way. Three important consequences follow: the wave is distorted from its square form, its speed of travel is reduced, and the circuit, instead of being almost wholly concerned with the *transmission* of energy, becomes involved also in energy *storage*.

It is the question of energy storage which ought primarily to engage our attention. Without any need for precise mathematical expression it can readily be appreciated that wave action becomes less free as the balance tilts away from the input of 'new' wave energy and towards more or less long-term energy storage in the medium of propagation. In our electrical example it would be quite easy to secure that the distant station should receive, for all practical purposes, no battery current at all, but merely the end-product of a long chain of capacitor charges and discharges, and we might then break the battery connection completely without causing any immediate cessation of signal at the far end of the line. It may be helpful to add a horological analogy. A clock pendulum or balance beats as nearly as practicable in a perfect sine-curve, though the driving force applied to it is an intermittent square wave-form. The drive is weak, the inertial mass large, and consequently it is the driven part of the system which almost completely determines the frequency and wave-form. But deprive the pendulum of its bob, and the bare rod will become much more responsive to the impulses it receives from the escapement: it will be forced beyond its 'natural' frequency of swing, and the reversals will be more sudden and jerky.

The crucial issue is evidently one of frequency. A heavy pendulum, a charged capacitor, or a coil with a magnetic field through it, is an energy-storing system, and its condition can be reversed only just so rapidly as the limited forces which can be applied to it will permit. A medium which absorbs and stores energy in this way consequently transmits waves differentially: low frequencies may pass where high ones do not. It is further to be noted that frequency is implicit in wave-form. The instantaneous rise of a square-wave crest is simply the counterpart of an infinitely high frequency, and in textbook introductions to Fourier analysis it is almost standard practice to show how a square wave can be resolved into its fundamental frequency and an infinite series of harmonics. Put a square wave into a medium with limited frequency response and the higher harmonics will be stripped off, possibly to the extent that the

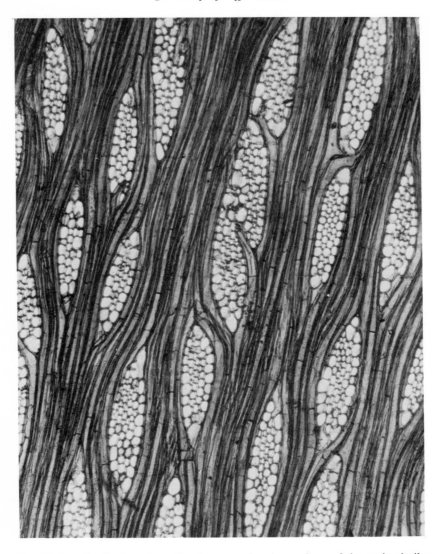

Fig. 40. Longitudinal section of mahogany, showing a slope of the grain similar
to that disclosed by splitting in Fig. 35 at p. 101. The predominant slope here may
be called Z, by reference to the middle stroke of that letter. At upper centre are
two configurations which look like intrusions of tissue with an opposing slant
across one of the very large rays which characterise this particular tree. If such an
intrusion succeeds it must be classed as an S-event, again by reference to the form
of the letter. Again at lower right there is a critical situation where Z-orientation
is being maintained by a single cell: death of that cell would be an S-event, whereas
its duplication would be a Z-event.

residual wave is hardly different from a pure sine-curve. Limitation of frequency is nearly related to limitation of velocity: which aspect is preferred for observation in a particular case can only be a matter of convenience.

Practical demonstration of these effects in differentiating tissues is laborious, but not specially difficult in any other respect. We may use examples related to the photographs already introduced (Figs. 32–35). The visible waves in these specimens owe their existence to a *double* inclination of the cambial cells in the tree. We saw at p. 103 and in Fig. 37 that these cells are sloped a little inwards at their top ends, but this was an incomplete account of the position. The cells in fact are sloped also in a plane at right angles, that is to say circumferentially round the trunk of the tree. It is useful here to refer to the capital letters S and Z. If these are inscribed on the outer face of a tree trunk it will be seen that their middle strokes are inclined in opposite directions. We may now say quite generally that mitotically active cells in the trunk are oscillating continuously between the S and Z orientations. In fact almost every incident which can occur relating to the division, growth, or death of a cambial cell is plainly and unambiguously classifiable *either* as a Z-event or as an S-event (Figs. 40 & 41). Furthermore there is a pronounced segregation of these categories into distinct areas or 'domains', the purity of the S and Z characteristics of the domains, and the sharpness of the boundaries between them, increasing as the tree matures. At a given moment, therefore, the cambial surface is a mosaic of S and Z domains. In some specimens the domains are almost stationary, but in others the whole mosaic moves vertically, so that by appropriate choice of laboratory subject we can apply to a given portion of tissue an almost square wave-train of S and Z histological events, with considerable freedom to regulate frequency and wavelength at will. The system also possesses, in a high degree, that selective opposition to high frequencies which we would represent in an electrical model by impedance and in a mechanical one by reciprocating inertial mass. That is to say, a domain of S events entering tissue with a pre-existing Z polarity will reverse that polarity to S only if the frequency is low enough. If the wave-train is moving too rapidly the Z polarity will be only partly neutralised before the S domain has passed on and the next Z domain has arrived to restore the *status quo*.

Our interest in this system lies essentially in its complete ordinariness. It has been extensively researched without yielding any relationship which creates even a momentary difficulty in the application of wave concepts. For example it is found (Bannan, 1966) that in common conifers the grain of the trunk becomes progressively S in early life and then swings back to a Z polarity which is never reversed. There are always both S and Z domains, but these are in constant motion and there is a shift in their

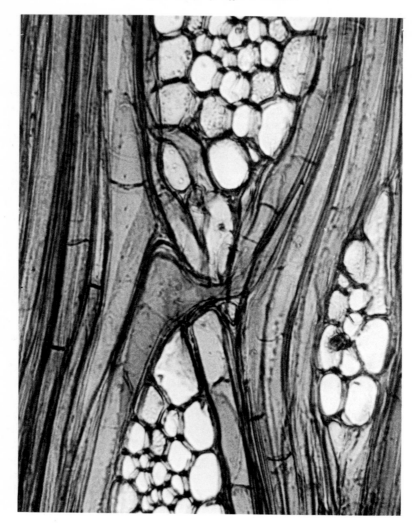

Fig. 41. Visually obvious conflict of cellular orientations in mahogany. The cambium of a tree (*every* tree, as far as we know) exhibits a multitude of such local conflicts, the resolution of which is biassed by a system of probability waves. The effect of isolated decisions upon the general slope of the grain is slight; to produce a conspicuous swing it is necessary that one orientation should gain the victory in a high proportion of cases and over a prolonged period of time.

relative proportions. What would constitute a mathematically congruent electrical model for this? Make a rotating commutator with adjustable sectors, initially set at equality, connect it to a battery, run it at speed, and feed the output to a voltmeter. If the needle is efficiently damped it will not move, but if the sectors are shifted so that one polarity is applied through a greater angle of rotation than the other, an average reading will appear. In the same way the polarity of the tissue can accurately reflect the general balance of S and Z domains at frequencies where it fails to register their impulses separately.

How can we experimentally determine the upper frequency-limit for the response of the tissue? Suppose that we run the domain wave-train progressively faster and faster and observe how well the structural polarisation of the wood manages to keep pace. As the frequency increases the response will fall into arrears by an increasing fraction of the cycle, and it is not difficult to perceive that a critical point will be reached when the delay amounts to a quarter-wave. Any further increase in frequency, and the imposition of a majority of S-events will actually come *before* the Z-polarisation has reached its peak. The impulses are now so badly mistimed that they are no longer stimulating the oscillation of the tissue, but actively suppressing it. Measurements of a specimen very near its critical frequency have been given by Krawczyszyn (1971, 1972) and compare instructively with observations by Pyszynski (1972) on another tree where the frequency was well below the limit.

In one respect only does the application of wave concepts in histology seem likely to go a little beyond the intuitive perceptions which may reasonably be expected of a biological student. The customary elementary treatments of wave interference are all somewhat artificially simplified by the assumption that the waves involved all travel at the same speed. This is legitimate for many optical purposes, and will go far enough in acoustics to explain why a wind instrument produces a note of particular musical pitch. But a wooden flute playing the same note as a metal one produces a different set of harmonics, and the cause lies in waves moving in the body of the instrument at speeds different from those prevailing inside its bore. In relation to S and Z domains it has already become necessary (Hejnowicz, 1973, 1974) to allow for interference between wave-trains having similar frequencies but very different speeds of advance. Such interference causes standing beats; that is to say the domains show a pulsation of size without much change of location. Hejnowicz modestly refers to his analysis as 'a hypothesis', but in reality it has to be treated as a fact of observation, and it presents us with a kind of mathematical framework which is automatically and instantaneously transferable to a multitude of other problems. Compound wave-trains of this type, radiating from appropriate centres, will for example very readily generate spotted or reticulate patterns in the style of giraffe or leopard skins.

7

Structural analysis of complex tissues

All mass-action phenomena of differentiation can be treated as waves, and because the mathematics of waves is thoroughly well understood, the purely computational problems are thereby finally disposed of. Wave concepts offer no remedy for the technical intractability of a specimen, and give little assistance in such matters as the formulation of appropriate biochemical hypotheses, but at least they can reduce the quantitative relationships between cause and effect to a mechanical operation of arithmetic. When we come to consider specific incidents at the cellular level (whether as elements in a wave or as items in some more diffuse developmental pattern) there is no corresponding general system of ideas ready for our use. We can only return to first principles, and make our own way.

We first introduced the steady state as a kind of datum level of histological differentiation, and at p. 90 we more formally equated the organisational attainment of a steady-state tissue to zero. It follows inescapably that we ought to ask what meaning, if any, can be attributed to such a statement as:

Level of histological differentiation attained $= 27$.

This enquiry leads easily into a fundamental theorem, which is simply that the structural complexity of tissues conforms to a natural whole-number scale, so that processes of differentiation are always subject to a kind of quantum principle, fractional advances in organisation being prohibited. If a tissue seems to differentiate gradually that is only because its level of complexity is already represented by a very large number, just as a chemical reaction may seen to be gradual because it involves many molecules which are not being observed individually.

In order to appreciate this truth we have to ask what is the irreducible and indivisible unit of observation in the structural complexity of a tissue. In putting cells together to form a tissue, what is the *basic mathematical element* of the process of assembly? Evidently, the juxtaposition of two cells, one against the other. We will call this a *contact*, and if we remember that a contact is two-sided we have it immediately from previous work (pp. 37–8 and 57–63) that in any continuous tissue the total number of contacts will be three times the number of cells as seen in section but seven times the number of cells when considered in the solid. An enlargement of the common meaning of the word is plainly implied; the description of a contact must be deemed to include every aspect of mutual relationship between its constituent cells. A contact is not to be seen as a mere interface, but as a biological association within which any style of behaviour may be displayed from active partnership at one end of the scale to unrestrained predation at the other.

It will be convenient to use capital letters to denote recognisable classes of cell, and to bracket two letters together to denote any class of contact. Suppose in any tissue we have (perhaps among a multitude of others) some cells which we recognise as members of a category J and others which we assign to class K. We now have the *possibility* of contact classes (JJ) (KK) (JK). Which, if any, of these kinds of contact actually exists in the tissue can only be determined by search. There is a reciprocity of relationship between cell- and contact-classes. Cell-types define contact-types, but contacts also enter into the description of cells. The question whether cells of type J have (JK) contacts, and if so with what degree of consistency, is part of the description of the class J, and indeed may quite conceivably come to figure as an essential item in the *definition* of that class.

The only way to understand the interplay of cells and contacts is by making trials with simple diagrams. It is enough to work with linked polygons analogous to tissue sections. First draw a figure in which all the contacts are identical. This proves to be a complete specification for a network of equal regular hexagons, and most students have no difficulty in seeing not only that this is a solution but also that it is the only solution. The next step is the production of a diagram in which there will be exactly two kinds of contact. The conceptual problems now become much more severe, and the average biologist will take a long time to explore the possibilities. In fact there are just four solutions:

(*a*) In a chess-board array of squares, open out every vertex into a 4-gon.

(*b*) In a net of regular hexagons, open out every vertex into a 3-gon.

(*c*) In a net of regular hexagons, insert a 2-gon in the middle of each side.

(*d*) In a net of regular hexagons, letter each cell J or K in such a way that every cell J has six (JK) contacts and every cell K has three contacts (JK) and three contacts (KK).

Here is the beginning of a fascinating mathematical game. We need not pursue it much further, but there are a few important lessons to be learnt even from these rudimentary examples. Take for instance solution (*d*) above. If we commit an irregularity in preparing this diagram what happens? We shall have a cell possessing a set of contacts which is neither that proper to a cell of class *J* nor in conformity with the definition of class *K*. Such a cell can only be correctly labelled by the introduction of a third letter, but this means that all its neighbours now have a new type of contact, and will themselves require to be reclassified, and in this way the consequences of one initial misplacement will spread through the tissue like an infection.

This is only one particular manifestation of the general rule that a very large number of contact-classes is required for any approach to general biological verisimilitude. As the diversity of contacts is limited by the diversity of cells we are obliged to dismiss as inaccurate a good many common habits of thought and speech. If it is said, for example, that a tissue 'consists of two types of cell', the suggestion needs to be evaluated by comparison with the four geometrical constructions we have just reviewed.

We have not yet completely resolved the question whether the level of organisational complexity is better represented by the number of contact classes or the number of cell classes. Our recent geometrical exercises do not directly help in this matter because they display (fortuitously) a coincidence of the two numbers throughout. We can however draw some useful guidance from the breakdown of attempts which may be made to circumvent the restrictions we have discovered. Why, for example, can we not contrive to have two kinds of contact with only one kind of cell? Why not squares in staggered rows (side-to-side and half-side-to-half-side) or elongated cells (side-to-side and end-to-end)? In the event, all such enterprises are frustrated by the emergence of three contact-classes, not two. We can arrange for two of the contact types to form a stereo-pair but we cannot make them identical. A pattern of equal regular hexagons has one cell-class and one contact-class, but a pattern of equal elongated hexagons has one cell-class and three contact-classes. The advance in organisation is here shown by contacts, not cells, and further exploration will be found to confirm the general soundness of the view that the only natural and proper measure of the structural complexity of a tissue is the number of contact-classes existing within it. We may notice here a strictly logical relationship between our theoretical development and the actual conduct of observational work in the laboratory. Suppose we have a specimen with many distinguishable cell-classes in which a lot of the theoretically possible contacts fail to materialise. We have, let us say, cells *J* and *K*, but no contacts (*JK*) or (*KK*), and similarly in other cases. What is the condition of such a tissue? Obviously there is a regularity of construction, an approach to diagrammatic standardisation, and our record-keeping is simplified accordingly. We can make no measurements on a situation which does not exist. But in a specimen of more varied constitution the 'missing' contact classes are likely to appear, and our notebook will expand as we struggle to estimate the frequency with which the (*JK*) type of intercellular confrontation will occur, and to assess the biological consequences when it does, and similarly for all the other combinations.

Contact diversity in the steady state

That two individual contacts should be absolutely identical is possible in a geometrical construction, but not a realistic expectation in even the simplest living tissue. A block of solid tissue consisting of 10 000 cells will contain about 70 000 contacts with not a single case of exact duplication. If this massive exhibition of diversity is duly recorded in our notebook our next task will be to partition the variability into three sections:

(*a*) Purely random deviations, of no scientific interest. Always present, and to be delimited by conventional statistical procedures.

(*b*) Diversity equivalent to that of the steady state. Necessarily inherent in all living tissues, but of general biological interest only, and incapable of characterising particular specimens.

(*c*) The residue (if any) after (*a*) and (*b*) have been subtracted from the total contact diversity of the specimen. Only this residual diversity has the status of 'differentiation' or the power to express taxonomic or genetic differences.

The crucial point in this scheme is the subtraction of steady-state contact diversity from total contact diversity. How is this operation to be performed? From earlier chapters it may well seem that this must be a very complex problem. We know that interfaces in the steady state display variation in polygonal grade and that some of them are new creations while others are not. To disentangle just exactly *this* amount of variation from a situation which is (to an unknown extent) of even *greater* intricacy perhaps appears at first glance to call for extraordinary mathematical skill.

In reality all these apparent difficulties are easily swept aside. We already possess the necessary mathematical tools; we shall simply take the principle that complexity is measured by the number of contact-classes, and extend its application from diagrams to real tissues. Writing our fundamental equation in its final form:

Level of differentiation = (Number of contact-classes) − 1

it becomes clear that the key to the analysis lies in the classification of contacts. We have merely to organise our work according to some logically coherent body of rules which will have the effect of assigning to one and the same class *all* the contacts of a tissue which perfectly exemplifies the steady state. But this is simply another way of saying that in order to qualify for membership of any contact-class additional to the first a contact must exhibit some feature which is not a passing incident of the mitotic cycle, but of a more persistent nature. In any complete analysis of a tissue the phenomena of anisotropy and polarity will be automatically comprehended within the list of contact-classes, but this is not a relationship which seems likely to find much laboratory application. The common form of research project will be a study involving permanent or semi-permanent differences between cells or, in botanical language, questions of idioblastic distinction.

Because some readers may feel hesitant about reducing all the contact diversity of the steady state within the limits of a single category it will be desirable to look more closely at what is involved. It must be appreciated that a contact is purely abstract. It is an association or relationship, not a physical structure capable of being viewed through a microscope. A familiar way of expressing relationships is by a two-way correlation table, so let us prepare a suitable grid, say 10 rows × 10 columns, and consider how the contacts of a steady-state tissue are to be arranged within it. Our first task will be to insert appropriate column-headings, and for this we can only use stages of the mitotic cycle: there is no independent alternative measurement available, and any attempt at evasion (e.g. the substitution of cell-volumes) will amount merely to a change of scale. Having headed our columns we must perforce distinguish our rows in an equivalent manner. Any individual contact now takes its characteristic place in the table as being, let us say, the association of one cell which is at metaphase with another which is seven-tenths of the way to its *next* metaphase.

We have next to consider what measurable properties can properly be attributed to a contact, rather than to a cell. Obviously such things as area of interface, polygonal grade of interface, probability that interface shall be a 6-gon. Obviously not cell-volume (but by all means, if desired, the combined volume of two cells). Many physiological possibilities such as membrane permeabilities and rates of hormone flow. Quite conceivably, for there is nothing in the steady state to prohibit individual variability, such matters as an influence by one cell upon the future mitotic activity of its neighbour. The list is extensive, but it is subject to a simple mathematical rule. For each place in our table each contact-property we choose to measure can have only a single average numerical value. A specially instructive case to consider will be that of cell-wall thickness because this will enable us to deal with the apparently intractable problem of incorporating into a common scheme the contacts which are associated with newly created mitotic partitions and those which are associated with older portions of cell surface.

Let us so arrange our table that the first compartment (that at top left of the grid) relates to contacts between two newly-formed daughter-cells. We may write the thickness of a mitotic partition at the moment of its formation as zero, but the entry in our table compartment must be rather more; no doubt the juxtaposition of two new cells will imply that the wall between them is also new in a high proportion of cases, but it will not always do so, because we must allow for the accidental meeting of new cells which are not sisters, and which are separated by an older and therefore thicker wall. Now let us move down the descending diagonal of the table until we come to the type of contact which involves two cells just about to go into prophase. Of course such cells may be sisters, and the wall between them may therefore be still not quite of its final thickness, but the distinctive effect of the mitotic partition is evidently diluted rather severely by the time this entry in the table is reached. We have therefore a characteristic distribution of averages: at the upper left corner of the table is a concentration of sub-standard wall thicknesses representing the inflow of newly created mitotic partitions. Down the diagonal the average thickness increases, probably rather rapidly, without however necessarily quite reaching the value which represents full maturity. And as the values on the diagonal are rather low, those remote from the diagonal must in compensation be a little higher than the *general* average for the whole tissue. The reader will easily visualise the cognate table for polygonal grade of interface, remembering that mitotic partitions are to be credited as 6-gons while all the table entries together must average out at 5.143.

It is in fact perfectly logical that mitotic partitions should be assimilated into the general population of contacts; this is simply a reflection of the phase-ambiguity of all periodic phenomena. An interface which is not a 'new' partition now must have been so at some time in the past. The difference between an old wall and a new creation is only a difference of mitotic phase, a difference possibly of some thousands of cycles, but still only a difference of phase and not a difference in anything else.

We have referred to a table of 10×10 entries, but finer subdivision of the mitotic cycle can be adopted as desired, and the effect will be to accentuate the peculiarities arising from the inflow of new mitotic products at a particular locality in the table. The theoretical conclusion is inexorable: the contacts in a steady-state tissue form in all respects a single continuum of variation. They exhibit diversity, but they will not admit of classification. They can be individually described, but there is no way of sorting them into categories which can be validly segregated one from another.

At this point the reader may be tempted to say that because a steady-state tissue contains separate categories of interface it 'must' contain separate categories of contact. In reality no such connection exists. No matter how carefully we draw up the definition of a category of interfaces, it will prove inadequate as a definition of any recognisable category of contacts. Between interface properties and contact properties there exists only a loose and imperfect correlation, and the resulting ambiguity is enough to permit continuous gradation of contacts and discontinuous classification of interfaces to coexist in the steady-state tissue.

For the time being it seems clear that the more exact determination of the distribution of contact-properties in a steady-state tissue is not a realistic proposition. No doubt something might be done to develop a mathematical formulation, but this would be pointless as there is evidently no prospect of obtaining laboratory facilities on the enormous scale which would be required for any worth-while attempt at observational verification. Nor would an operation of this kind have any claim to major biological importance. We need to be fully aware of the existence of a continuous distribution, but we do not urgently require detailed information about its form.

Idioblastic classification of cells

It was convenient in the preceding section to associate measurable properties of a tissue directly with contact-types. In laboratory work it is hardly ever possible to follow this course, because the cell possesses an unshakable supremacy as the natural psychological unit of observation. Normal routine will therefore be indirect: we shall measure cells, but the cells we measure will be chosen because of the contacts they possess. This can be expressed algebraically by means of a *bipartite symbol* thus: $(t)K$. Here the capital letter denotes a cell-class, the bracketed lower-case letter some measurable property of that class. Any bipartite symbol will be validly constituted so long as its cell-class is defined idioblastically (i.e. in terms of contact-properties more permanent in nature than those of the steady state) and its measurable property has a single average numerical value (it need not be a direct observational result: such quantities as a probability or a standard deviation will be admissible).

We are now in a position to see how the laboratory investigation of a complex tissue must be organised. If we could obtain a complete list of numerical values for all the bipartite symbols which can be correctly constructed according to the rules just stated, then our work would be finished. We would have transferred to paper (within the limits of sampling error) everything which it is possible to ascertain concerning the differentiation of the tissue. This is an unattainable objective, so we can only proceed selectively. Given a prior interest in some problem of development, we must think first what types of contact are likely to produce results which may be informative upon our chosen topic. From this stage we progress to formal definition of some relevant cell-classes, and to a consideration of the kind of measurement which it may be appropriate and possible for us to make upon those cells. In this way we arrive at the formulation of a list of bipartite symbols (if we are wise it will be a *short* list), the values of which we may hope to obtain by observation. The next stage of the operation is to find the required cells.

There is a fundamental natural law, familiar to every research worker, which decrees that any phenomenon which it is particularly desirable to observe shall be of very infrequent occurrence. This principle applies with devastating force to the study of tissue differentiation. We cannot ordinarily expect, even by micromanipulation or tissue culture, to exercise precise control over the juxtaposition of cell against cell. Structural configurations which may be crucial to our understanding cannot be contrived at will, but must be located by search operations which will often range over large areas of tissue in which there are only common types of cell in which we have no specific interest. For this reason advanced histological analysis is restricted, for the foreseeable future, to tissues in which there is some conspicuously distinct form of cell, preferably isolated, which will serve as a signpost or marker. The tissue region surrounding such a cell is our counterpart of the physicist's cloud chamber or cosmic-ray detector. It may not be, probably is not, a place where interesting events are specially likely to occur, but it is a place with special facilities for detection and measurement, and an investigator who turns away from these advantages deserves to be censured for inefficient use of resources. An unselective observational programme, in which cells are recorded indiscriminately just as they are encountered by chance (at the worst, *every* cell in a compact block of tissue) fails because the enquiry is in fact of a composite nature. A laboratory working in this way is in reality conducting a large number of research projects concurrently, many of them highly disparate, switching manpower and instrumentation continually from one topic to another.

An idioblastic classification of cells follows the common taxonomic pattern in being thoroughly hierarchical in structure. We have repeatedly the option to subdivide cell-classes, or not, according to the needs of the moment. A universally applicable system of symbols is out of the question because the resources of the alphabet are so restricted. It is much more a question of improvisation to meet the requirements of a particular enquiry, and even in quite a limited field one cannot avoid the use of distinguishing subscripts and other rather unsatisfactory devices.

For the purposes of some worked examples given later in this chapter the chosen marker is the stoma of a plant epidermis. (By disregarding the final mitosis which produces the two guard cells we are merely referring back to a stage of development a little earlier than the one commonly observed). We may therefore use S as a symbol for the class of all stomatal mother-cells. Recognition of this class creates a consequential classification of the surrounding cells. We have a good deal of discretion in the treatment of this classification. Any logically consistent arrangement will be equally correct with any other; the ultimate units of the taxonomic hierarchy are geometrically prescribed for us, but the framework of larger groupings is at our own choice, and the considerations governing that choice are practical, not theoretical. We require a classification which will be a serviceable laboratory instrument, which means above all that the major cell-classes must be instantly recognisable. Let us institute three more classes:

A (for adjacent): to include all cells with one (*AS*) contact but no more.
B (for between): to include all cells with two or more (*BS*) contacts.
R (for remote): to include all cells without any (*RS*) contact at all.

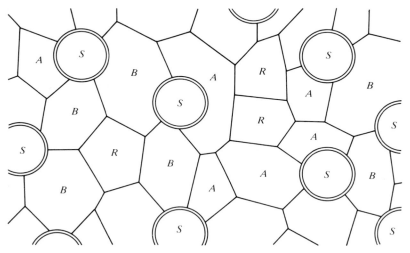

Fig. 42. Recognition of primary cell-classes. The emergence of an idioblastic category *S* automatically distinguishes the other cells into classes *A*, *B* and *R*. Subordinate classification within these primary divisions can be arranged in a great variety of ways according to the direction of our interests: for example, the cells of class *A* in this tissue might divide into three subclasses according to their possession of (*AB*) contacts (one such contact, two, or none), but for other purposes the same cells could be re-classified on the basis of their (*AR*) contacts. All the other classes will similarly admit of variant classifications according to the object in view.

We have now a system in which any cell is a member of one, but only one, of the four classes *S*, *A*, *B* or *R* (Fig. 42).

A scheme of this kind immediately generates a considerable array of hypotheses and questions about the action of cell upon cell. Suppose simply that some influence emanating from *S* changes the behaviour of the surrounding tissue. We have at once such problems as:

(*a*) Does the class *B* show the effects of double or treble dosage?

(*b*) If so what is the principle of combination – additive, super-additive, or sub-additive?

(*c*) Is there secondary leakage from the first cells affected, or does the influence of *S* merely lie where it first falls?

All such questions must be approached by refining the system of cellular classification. For example under question (*c*) above it will become an issue whether a cell of class *B*, having received perhaps a treble dose of the hypothetical *S*-influence, then passes some of the excess to a neighbour belonging to class *A*. For this kind of enquiry, therefore, we shall require to subdivide class *A*, at least to the extent of separating cells with (*AB*) contacts from those without. By similar but not identical reasoning we may be induced also to contemplate other perfectly specific types of laboratory project: an attempt, for example, to find cells *R* with as many, and as few, (*BR*) contacts as possible.

When arguments of this kind are energetically developed, the *SABR* primary classification proliferates endlessly into an ever-increasing number of more or less independent lines of enquiry. There are only two considerations which appreciably restrict our freedom of action. The first of these is the need to find a laboratory subject which is constitutionally adapted to a particular theoretical interest. If the cells of the class *S* are so close together as to make the class *R* non-existent, or too far apart for any cell to belong to class *B*, the tissue is less attractive, from the research standpoint, than one with a nicely-balanced mixture of cells. A less obvious point arises from the subdivision of classes. As we stipulate more exactly the conditions in which we desire to take an observation, so the cells in which we are interested become less abundant, and the proportion of the tissue which we are simply to discard as irrelevant rubbish approximates ever more closely to 100%. In normal laboratory conditions this generates a perfectly clear-cut statistical result, in that the effective estimation of cell-class properties can easily be continued far beyond the practical limit of estimation of cell-class frequencies. The basis of this distinction is psychological. If we instruct an observer *X* to search sections under the microscope and to measure the area of every seven-sided cell he happens to discover which has exactly two five-sided neighbours, his assignment is tedious but bearable. The period of continuous mental concentration which is demanded of *X* is only that sufficient to cover the scrutiny of eight cells and area measurement of one of them. For a great part of the time *X* can afford to relax his vigilance to a comfortable level: he is told at the outset that he is not to be held responsible for missed opportunities. But consider the situation of another observer *Y*, who is told to examine sections and identify *every* cell which is eligible for area measurement by

X. Although the mere identification of a cell-class is a simpler procedure than area measurement, *Y* has a much harder task than *X*, because it permits no relaxation of attention. In all comparisons of this general type, *X* will be found to have far greater endurance than *Y*, so that the properties of cell-classes are accessible to investigation where their frequencies are not.

Examples of cellular interaction

In the preceding pages we have established a set of design parameters sufficient to guide us towards reasonably efficient standards of planning in our laboratory work. In the last analysis we have done no more than add precision to the idea that research objectives should be clearly conceived in advance, and limited in ways which will make them realistically attainable. Existing publications upon the histology of differentiation however have not generally been designed around the intensive quantitative study of any particular thing. The predominant tradition is one of generalised literary description, to which processes of measurement may be added as a kind of optional embellishment. Many of the tissues examined display structural complexities going far beyond those of the basic *SABR* system of cellular classification (which in itself would seem to offer a man plenty to do). Extensive series of observations have sometimes been recorded with very great labour, but with astonishingly little preliminary consideration of any stated purpose, and even where an author has hit upon something of scientific value it is common to find that his extraction of results is incomplete.

A good example can be found in the work of Bannan (1966, 1968), who determined mortality rates for the products of pseudotransverse divisions in the cambial cells of conifers. Omitting various complications which are peculiar to this specialised field of botany, the situation here is simply that an elongated cell divides into upper and lower derivatives, both of which then give rise to cell-lineages of varying extent and persistence. Records were kept for no less than 68 000 initial divisions in a considerable range of species. Within a standardised test period there may be extinction of the upper cell-lineage, or the lower, or both, or neither. Of all biological distinctions, that between life and death would appear to be one of the most important, and for any living cell it would seem to be a matter of some consequence whether its immediate neighbour is alive or dead. We have therefore an obvious principle of contact-classification, but no recognition of this can be found in the original texts. The question whether a cell's own expectation of life might be affected by the death of its neighbour was not one of the issues which led to this very large observational assignment being undertaken, and was not among the problems considered worthy of discussion at the conclusion of the work. Yet it was plainly implicit in the

Table 8. *Survival of cell lineages arising from pseudotransverse division of fusiform initials in conifers. Calculated from data of Bannan (1966, 1968)*

	Whether upper or lower more likely to survive	Whether extinction of one favourable to survival of other
Cupressaceae		
Thuja occidentalis	Lower	No
Libocedrus decurrens	Lower	No
Cupressus sargentii	Lower	No
Other conifers		
Pinus contorta	Upper	Yes
ponderosa	Upper	Yes
banksiana	Upper	Yes
resinosa	Upper	Yes
lambertiana	Upper	Yes
strobus	Upper	Yes
flexilis	Upper	Yes
Picea glauca	Upper	Yes
engelmanni	Upper	Yes
mariana	Upper	Yes
Larix laricina	Upper	Yes
Pseudotsuga menziesii	Upper	Yes

whole proceeding, and a simple calculation yields results of a kind not imagined by the observer himself. Taking for purposes of demonstration the data on *Pinus ponderosa*, we have failure of one lineage in 57.1% of divisions, survival of both lineages in 27.9%, double failure in 15.0%, mortality rate for cells consequently $\frac{1}{2}\{57.1+(2\times15.0)\} = 43.55\%$. But if the fate of a cell were independent of the fate of its sister, the rate of double mortality would then be $43.55^2/100 = 18.97\%$, which is 3.97% in excess of what is observed. In this species, therefore, we find an indication that some effect is reducing the incidence of double mortality; the death of one of the products of a pseudotransverse division is contributing positively to the survival of the other product. Carrying the same form of calculation through the whole mass of data, and combining the results with records on the preferential survival of upper or lower product respectively, we find the taxonomically interesting situation displayed in Table 8.

Even the most isolated of scientific discoveries must obviously have some value, but if we are to develop any comprehensive understanding of cellular interactions it will not be satisfactory to rely on scraps of information recovered, by a kind of salvage operation, from investigations which were planned and executed without specific reference to our needs. The questions which we began to formulate at p. 122 are not necessarily very difficult, but they are not going to be answered by accident. To make any real headway towards a general system of accounting for histological special-isation, we shall require a laboratory programme based, *as an absolute minimum*, on some version of the *SABR* classification of cells, and on a consistent use of bipartite symbols.

It is easy to see from our previous work that special facilities are available if we choose to begin our operations by studying the polygonal grades of cells in surface view or in sections. There is a general law of six-side average, so any cell-class with an average polygonal grade other than 6 must be balanced with some other class or classes in the tissue. The sides of cells are thus analogous to units of currency, and the tissue must display a precise accountancy in which gains and losses (above and below the 'normal' level of six sides per cell) are always exactly equal. The attrac-tiveness of this idea is enhanced by the practical point that it is usually quicker to count than to measure, and by the further realisation that polygonal grade is likely to correlate with other important dimensions such as cell-volume.

A system of intercellular exchange

If we have a primary idioblastic cell-class S and use n to denote polygonal grade, the average polygonal grade for randomly chosen idioblasts will be written as $(n)S$ and we can easily find tissue samples in which this is very different from 6. Because $(n)S$ is defined as an average it will normally be fractional, but for any single idioblast the polygonal grade can only be a whole number.

Consider now the particular case of an idioblast which happens to be a 2-gon. This is a deficit of four sides, which must be compensated. There has been an outflow of four sides into the surrounding tissue. But a 2-gon cell has only two immediate neighbours. If therefore the outflow were equally divided between the cells which stood in a position to receive it, any cell which was next to a 2-gon idioblast would receive an influx of two sides extra to its own basic entitlement of six.

Evidently we need to classify those cells which are neighbours of members of the class S according to polygonal grades of the idioblasts. Let us employ numerical subscripts (Fig. 43), writing:

$(n)A_3 =$ (average polygonal grade for randomly chosen cells of the class A which are adjacent each to one 3-gon member of class S);

$(n)B_{3,5} =$ (average polygonal grade for randomly chosen cells of the class B, each situated between two cells of class S one of which is a 3-gon and the other a 5-gon).

These are properly defined bipartite symbols specifying measurable properties of recognisable classes of cell, and they are only two examples from a list of quantities which is potentially capable of further extension. How many of the specified averages will admit of practical laboratory estimation can only be discovered by trial: it is easy to write such an expression as $(n)B_{2,2,4}$ and the meaning of the symbol is self-evident, but there is no

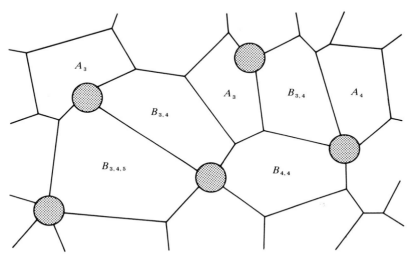

Fig. 43. Subdivision of cell-classes A and B by attachment of number subscripts denoting polygonal grades of adjoining idioblasts (shaded). This subdivision is a convenient one for studying the balance of polygonal grades in the tissue, and enables us to observe the relationships shown in Fig. 44. Even for this particular purpose, however, the scheme of subscripts is a matter of choice, and it is quite conceivable that alternative classifications might also give rise to successful programmes of laboratory work.

way of knowing beforehand whether the specified cell-class will occur in a specimen with a usable frequency, or indeed occur at all.

Adhering to our original notion that outflow from an idioblast may be equally apportioned among its neighbours the influx experienced by any A or B subclass can be calculated directly from the distinguishing subscripts. *On the hypothesis* that there is no secondary transference of sides, so that influx simply lies where it first falls, and remembering that each cell class must be allowed six sides in its own right, we at once obtain a complete set of predicted values:

$$(n)A_2 = 8.0 \qquad (n)B_{2,2} = 10.0$$
$$(n)A_3 = 7.0 \qquad (n)B_{2,3} = 9.0$$
$$(n)A_4 = 6.5 \qquad (n)B_{2,3,4} = 9.5$$
$$(n)A_5 = 6.2 \qquad (n)B_{3,5} = 7.2$$

and so on. Our hypothetical basis for these calculations is quite impossibly rarefied. Exact conformity of any real tissue to this sequence of numbers is not in the least probable (which is just as well, because such conformity would be singularly dull and uninteresting). The real value of our procedure lies in the assignment of numerical values to our concept of influx: we can calculate, from a cell's position in the tissue and its experiences of contact with other cells, something which it is reasonable to see as a kind of stress, something which is imposed by external forces and which will tend to drive the cell to a higher polygonal grade than the normal average of 6. Influx, in other words, may stand as the independent variable for a graph. For the dependent variable we shall naturally take the observed excess of polygonal grade above 6.

At this stage in the argument it becomes possible for the first time to see a whole complex range of structural relationships as a single great experiment. In elementary physics teaching we may stretch a wire by progressively adding weights, and plot a graph of strain against stress, and the slope of the line will give us Young's modulus for the wire. The histological analogy is precise. Our calculated theoretical values of the influx play the same part as successive loadings of the wire, and by examining cell subclasses in appropriate sequence we shall be able to see how the cell reacts to the increasing force which is being applied to it. For the measured stretching of the wire beyond its original (unloaded) length we shall in our own graphs have to substitute the observed excess of polygonal grade over the general datum level of six sides per cell.

The practical conduct of such a test is a perfectly simple matter, hardly exceeding the scope of a class exercise, so long as we use a reasonably convenient tissue and restrict our observations to the commoner classes of cell. A worked example is supplied from the leaf epidermis of *Plantago major*, treating the stoma (that is, the two guard cells taken together) as the primary idioblastic class S. In this tissue $(n)S$ is roughly equal to 3, and there is no technical difficulty of any kind in estimating mean polygonal grades for several subclasses of A and several two-subscript divisions of class B. The resulting graphs (Fig. 44) are amply sufficient to establish the existence in these leaves of a system of natural law. In both of the cell-classes A and B, increasing theoretical influx of sides induces a steady progressive increase in the observable excess. But as between these two major classes there is a very clear demonstration of a super-additive principle. Cells of class B, receiving influxes from two stomatal contacts, are driven to a polygonal grade substantially higher than strict arithmetical summation of the two influxes would lead us to expect. The tissue fortunately contains, though only in very small numbers, cells of class B which touch three stomata. With three subscripts the number of subclass combinations is rather large, and each of them, taken individually, is a considerable rarity. To prepare a complete graph for three-subscript B would, therefore, be a very extended enquiry, with most of the observer's time wasted in unproductive scanning of useless material. It has been possible however to combine data from different three-subscript categories and so produce a single graph-point. The line must pass somewhere near this point. Its slope is conjectural, but the super-additive rule seems to be extended in a consistent manner.

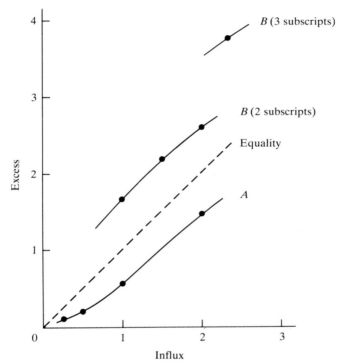

Fig. 44. Graphs of observed excess against calculated influx for cell-classes A and B, from original observations on leaf epidermis in *Plantago major*.

Other species of *Plantago*, some of them with significantly different general levels of $(n)S$, appear to be organised in a roughly similar manner, and in order to complete the picture we may note that $(n)R$ in these leaves is everywhere about 6.10. We have therefore a rather clear pattern of behaviour emerging. The stomata are in a general state of deficit, so far as their polygonal status is concerned. The counterbalancing excess of sides in the rest of the tissue is absorbed selectively by cells in contact with *more than one* stoma. Cells in contact with one stoma only also exhibit a certain excess of sides, but as this excess is normally *less* than the influx which would be due to these cells on a system of equal shares it seems that some quality of resistance or reluctance must be attributed to them. Lastly we have to recognise a low and apparently almost constant rate of 'leakage' of sides into cells more remote from direct stomatal influence.

Speculation about the biological significance of this geometry would evidently go beyond the scope of a book which is intended essentially as a compendium of methods. From the design of our procedure there cannot however be any doubt that the particular pattern of equalisation which we have discovered arises from a specific decision-making process in the development of the organism. There is no principle of mathematics from which anybody could have predicted the distribution, and it is easy to draw plans for tissues in which the whole matter would be arranged quite differently. We do not know very much of the reasons which induce a living tissue to settle down in one particular state of equilibrium and reject so many others, but by repetitive application of our scheme of bipartite symbols we shall have before us an absolutely unlimited prospect of empirical discovery, and we can direct our attention, and our technical resources, to any aspect of tissue behaviour we may wish to consider in detail.

To illustrate the management of a more specialised enquiry we will take the question of a possible reaction of the cell-classes *A B R* against the primary idioblasts of the class *S*. In a plant epidermis there are two conceivable states as follows:

(*a*) The stomatal mother cells, perhaps acting in concert, are in a position to impose their will upon the rest of the tissue. If a number of stomata in a particular locality choose to adopt lower-than-average polygonal grades, their neighbours have no option but to accept the polygonal upgrading which is thereby forced upon them.

(*b*) The non-stomatal members of the tissue are in a position to operate a quota system and set a limit to the total amount of upgrading which they are jointly prepared to undergo. Consequently any stoma which chooses to develop as a 2-gon will provoke a reaction, and other stomata in the vicinity will be restrained from adopting a similar course.

This is a perfectly clear-cut issue. It is a question about the seat of power, and a definite answer must exist; a decision cannot be taken in more than one place.

Appropriate observational procedure depends simply on the development of a system of subscripts for *S*. Working in tissues where the (*SS*) contact does not exist, we base our subscripts on the nearest possible approach between two members of the class *S*, that is to say on the confrontation of two stomata across a cell of the class *B* (Figs. 45 and 46).

S will take one subscript for each such confrontation, and to complete the scheme it would be necessary to write S_0 for the subclass of S without (SB) contacts. The number of subclasses of S, if it is possible to accumulate three or four subscripts, may be considerable, and some of them will be too

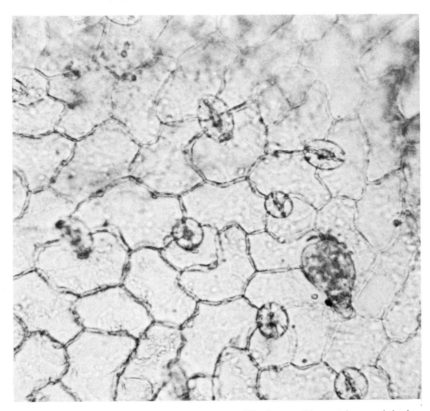

Fig. 45. Foliar epidermis of *Plantago major*. The large object at lower right is a hair: the presence of such structures, combined with unavoidable deficiencies of preparation, would make it extremely difficult to classify *every* cell across any considerable extent of tissue. Adopting a selective system of observation it is quite easy to find *isolated* examples of any desired cell-type which is reasonably frequent.

infrequent for their separate assessment to be very attractive as a laboratory assignment. It must be noted also that our problem cannot be solved by finding any single quantity such as $(n)S_{3,4}$. Our interest lies specifically in the differences between two or more subclasses of S.

Various observational strategies are possible. One might proceed, for example, by recording stomata indiscriminately, and then undertaking a regression analysis to find out how the polygonal grade of the 'central' stoma related to the sum of the influxes corresponding to its subscripts. Thus a stoma of class S_3 is exposed to only one cell of class B which is itself subject to an influx of one, whereas a stoma of class $S_{2,2,3}$ is exposed to cells suffering a combined influx of five, so if B has any power of reaction

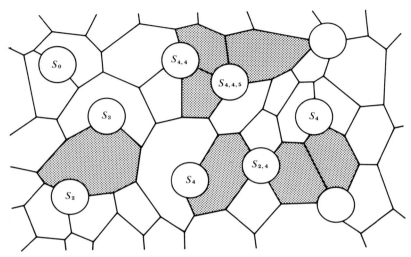

Fig. 46. Subdivision of cell-class S. Each idioblast is given as subscripts the polygonal grades of those other idioblasts which it confronts across intervening cells of class B (shaded). This system provides a basis for laboratory investigation directed to the question of possible reaction, by the rest of the tissue, upon the class S (see text p. 130).

the effect might be very considerable. No comprehensive investigation along these lines has yet taken place, but we have results from a simplified study. The material was leaf epidermis of a genetically uniform sample of *Plantago lanceolata* and the method adopted was merely the recording of polygonal grades for the three commonest single-subscript subclasses of S. We have approximately:

$$(n)S_2 = 2.650, \quad (n)S_3 = 2.862, \quad (n)S_4 = 3.006$$

and by normal statistical standards it would be necessary to accept the ascending gradient of these figures as genuine. In reality it would not be safe to rely upon a single trial, and it is unfortunate, to say the least, that the series includes no determination of $(n)S_0$. But if for a moment we take the data at face value, then the message about conditions in the tissue is perfectly plain. The stomatal mother-cells are acting, to some extent, in co-ordination, and the rest of the tissue has only very imperfect means of resistance.

This may be thrown into an almost dramatic light by contemplating the situation of a cell of class *B* which happens to be next to a 2-gon stoma. From previous work, we know that cells of class *B* already suffer more than their fair share of polygonal upgrading, and contact with a 2-gon stoma ensures that our chosen cell will be subject to maximum influx at one end. Is the cell permitted to secure relief by exercising restraint upon the stoma at its other end? On the contrary, the fact that one stoma is a 2-gon only makes it more likely that the other will be a 2-gon also. Furthermore the characteristic position of class *B* makes it only too probable that our cell must be the helpless carrier of those communications which must pass between the two stomata if their actions are to be co-ordinated in this way.

No doubt there will be those who will consider that this is too anthropomorphic an account, but there are some advantages to be gained by adopting a lively style of description and recognising the existence of conflicting interests among the cells. If it is said, for instance, that increased polygonal upgrading is no special hardship to a cell, there is nothing in the least frivolous about the obvious retort that the only competent judge of that question will be another cell, and that the members of classes *A* and *R*, from their relatively sheltered positions in the tissue, display a most conspicuous unwillingness to have extra sides thrust upon them. Unless we are to understand differentiation in some such terms as these, it is a little difficult to see how we are to understand it at all. To some extent the imputation of motives and purposes to the cells is inescapably necessary, if only as a stimulus to scientific curiosity.

A point which may trouble some readers is that the exchange of sides is almost purely notional, in that cells are formed as 3-gons, 4-gons, etc., and do not subsequently change their shapes. In a plant epidermis the distribution of excesses and deficits is determined almost entirely by the placing and orientation of mitotic partitions, with hardly any rearrangement of cells once formed. We have used a form of description in which it is said that cells of the class *R* accept very few excess sides, but an observer watching the progress of events would be more likely to express this by saying that these cells are disfavoured by their neighbours as anchorage-points for mitotic partitions. There may perhaps be biologists who are disposed to argue that only the second description is 'true' and that the whole conception of an exchange of sides is to be dismissed as fantasy.

More careful logical analysis seems to show that there is no issue here beyond a simple question of priorities and relative values. Traditional biological literature is based on the ideal of a 'picture in words', of total conformity to the visual image. If our principal aim is to describe a specimen in such a way as to ensure that no photograph of the organism will ever be needed, and if it is considered that a certain lack of intellectual clarity is a fair price to pay for progress towards the chosen objective, then the whole idea of a geometrical balance in the tissue becomes, in all probability, a useless impediment which should be jettisoned forthwith. But if one is attempting to reach the standards of the more progressive sciences and reduce all the visual phenomena to a system of algebra, then a certain brevity of expression has to be tolerated if the work is to proceed at all. A universal terminology to suit both styles of enquiry seems to be out of the question: the functional efficiency of vernacular languages is too low for such a scheme to be practicable.

A worked example in taxonomy

Towards the further development of our subject there are two steps which plainly ought to be taken without delay. In the first place there is an urgent need to extend our range of bipartite symbols by introducing measurable quantities other than n. More importantly it has to be recognised that if tissue geometry is to take its rightful place among the biological sciences it will have to transcend its problems of internal organisation and start making acceptable contributions to genetics, taxonomy and physiology. For demonstration we will use a geometrical technique to separate genotypic from phenotypic diversity in a variable species. There is no special intrinsic difficulty about applying geometrical methods of analysis to general biological purposes, and the work which follows is of the simplest possible kind; with a judicious choice of specimens the whole matter would reduce to a student exercise.

A convenient laboratory subject is the foliar epidermis of *Plantago lanceolata*. This is an immensely variable plant with an almost constant pattern of biparental inheritance. No two individuals are even approximately alike, and clonal cultures can be established, and their purity maintained, by ordinary horticultural methods. If care is taken to obtain material from geographically scattered sources, then one soon has a collection of plants which differ, in their own way, quite as remarkably as the domestic breeds of dog or rabbit. The only real drawback to this species is that although the plant is correctly listed as a perennial, most clones in field cultivation show a progressive loss of vigour and become unserviceable after a working life which in most cases exceeds five years but usually will not extend to ten.

We need consider only one primary idioblastic cell-class S, represented as before by stomatal mother-cells. The polygonal grades for S range from 2 to (very rarely) 7, and the laboratory work utilises a corresponding set of mechanical counters, upon which randomly chosen stomata from one leaf can be tallied to a total number normally in the range 250–350. The

Table 9. *Phenotypic variation in polygonal grade of stomata*
in Plantago lanceolata (*original observations*)

Average polygonal grade was separately determined for each of twenty
leaves of a clonal culture. Only the six most extreme leaves are shown
here

Averages	Counts of stomatal types			
	2-gon	3-gon	4-gon	5-gon
2.462	175	130	7	0
2.469	125	70	14	0
2.476	173	134	8	0
2.821	69	153	23	1
2.919	50	156	27	2
2.972	64	174	48	4

row of numbers standing on the counter dials after each leaf has been dealt
with provides not only a general estimate of $(n)S$ but also a more detailed
analysis of the stomatal population. When this procedure is applied to
leaves from any single clone it soon becomes apparent that individual
leaves can be markedly different, and statistical tests begin to show that
the differences cannot be dismissed as accidental. The general character
of the observations in genetically homogeneous material is illustrated in
Table 9. It is always a risky business to state empirical findings about the
extremes of a range, but it seems that for *Plantago lanceolata* as a whole
the value of $(n)S$ must vary at least from 2.25 to 2.95, and that at least half
this range of variation will quite commonly be displayed within a single
clone. Undoubtedly there are some clones which tend towards high values
of $(n)S$ and others which tend towards low values, and if these qualities
are anywhere combined with suitable restrictions on the range of phenotypic variation, then strains of the species may well exist between which
there will be no overlap, so that any reasonably accurate estimate of $(n)S$
will sharply separate one from the other. But it is quite clear that such a
separation would only be achieved in specially favourable circumstances,
and that the determination of $(n)S$ is perfectly hopeless as a *general* method
for making genetical distinctions in this material.

It is a point of some importance in botany to surmount this obstacle
if we can, because it is more or less standard practice to use the polygonal
grade of stomata as a taxonomic character wherever possible. The
variability which defeats the conventional method in *Plantago* is not
different in principle from that which causes difficulty in other groups of
plants, though it seems to be greater in amount than one would commonly

encounter. That the variation of $(n)S$ within a clone derives from environmental influence does not seem to be seriously open to question. Although not fully understood, it certainly relates to such matters as the

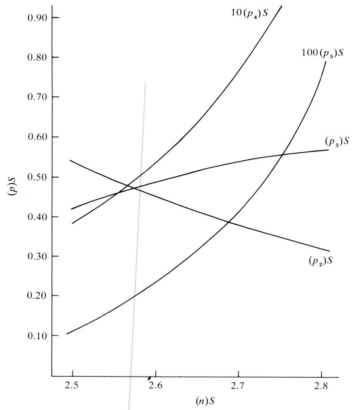

Fig. 47. Analysis of population structure in an idioblastic cell-class, from original observations on a clonal culture of *Plantago lanceolata*. $(n)S$ denotes average polygonal grade of all the stomata of any single leaf, and is subject to extensive phenotypic variation. $(p)S$ denotes the probability that any randomly chosen stoma shall be of the polygonal grade indicated by the subscript of p. Specimen analysis: in conditions where stomata of this genotype are to average 2.7 sides, a sample of 1000 stomata will contain roughly

$$(380 \times 2\text{-gons}) + (540 \times 3\text{-gons}) + (76 \times 4\text{-gons}) + (4 \times 5\text{-gons}).$$

passage of the seasons and the status of the leaf (i.e. whether on a main stem or on a branch), and by changing the horticultural practices one can somewhat imperfectly raise specimens to display high or low $(n)S$ values to order. This could be interesting physiologically, but our enquiry might appear to be running into a dead end from the taxonomic and genetical point of view.

We can however neatly evade a negative conclusion merely by inventing a new set of bipartite symbols. Let us write $(p_3)S$ as the probability that any randomly chosen stoma shall be a 3-gon, and similarly for $(p_4)S$ etc. Each of these quantities is necessarily a function of $(n)S$. By accumulating observations from any genetically uniform stock of material and using ordinary curve-smoothing methods it is a simple matter to obtain curves for the commoner grades of stomata (Fig. 47). From such a chart one can read off an analysis of the stomatal population for any specified value of $(n)S$, subject inevitably to the statistical restriction that only the middle portion of the range can be plotted to acceptable standards of accuracy.

It proves to be very generally the case that genetically different stocks have different idioblastic population structures for the same $(n)S$ value. In extreme cases this quite quickly becomes apparent to the observer. One culture has, comparatively speaking, a wealth of 2-gon and 4-gon stomata, whereas in another these types are scarce, and the 3-gon consequently much nearer to exclusive dominance, yet the *average* grade in both samples is likely to be 2.7 or thereabouts. Because the probabilities are not fully independent, but merely equipoised about the average, a single comparative chart will adequately display the taxonomic situation; we have however a choice, which can only be a question of convenience and taste, concerning the particular probability to be employed for the purpose.

Fig. 48 is in fact based upon $(p_4)S$, but as this quantity increases rather steeply with $(n)S$ it is necessary to make a suitable adjustment of graphical scale if the phenomena are to be nicely displayed upon the page. Therefore, instead of plotting the full value of $(p_4)S$ we plot only its difference (up or down) from a convenient linear function of $(n)S$. Of course this does not change the existing relationships among the plants: we merely substitute a sloping datum line for a horizontal one in order to space out the points more comfortably.

This operation apparently reduces the whole variability of the species, both phenotypic and genotypic, to one mathematically homogeneous family of curves. Such a situation must always be somewhat precarious, in that the very next observation may reveal a plant with a discordant pattern of behaviour, but for the time being we must take it that the geometrical determination of the epidermis in *Plantago lanceolata* proceeds upon the following basis:

(*a*) Every individual inherits a set of curves analogous to, but not identical with, those of Fig. 47, prescribing the exact proportional composition of stomatal population for every possible *average* polygonal grade of stomata.

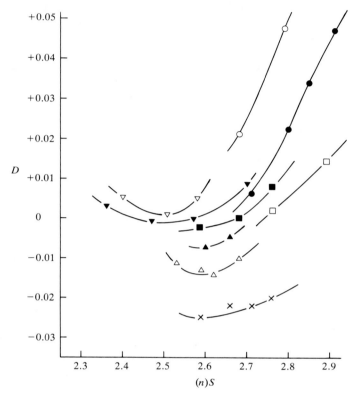

Fig. 48. Genetic differences in an idioblastic cell-class, from original observations on the proportions of 4-gon stomata in nine clonal cultures of *Plantago lanceolata*. Because a direct plot of $(p_4)S$ would be inconveniently steep the graph shows only a difference D, defined by:

$$D = (p_4)S - 0.171428\,(n)S + 0.394285.$$

Movement along any curve can be accomplished by phenotypic variation, but movement from one curve to another requires a change of genotype. Sample reading: when $(n)S$ is 2.75 the percentage of 4-gon stomata will be about 5.6 in the 'lowest' of these clones and about 11.5 in the 'highest'.

(*b*) In respect of this inheritance, the different genotypes stand in a linear sequence. That is to say, a plant which is intermediate between two others for one of its inherited curves will be intermediate between them for its other curves also. For example, a decision to produce an idiosyncratic excess of 5-gon stomata rather than the due proportion of 4-gons, thereby creating a population-structure which stood out of the main sequence, apparently would not be permitted.

(*c*) Every plant has considerable freedom to make local alterations in the average polygonal grade of its stomata. Such modifications are the subject of a secondary level of genetical control, to the extent that any genotype is likely to show a loose general predisposition towards low or high averages, but much of the variation is environmentally determined.

(*d*) The local average once fixed for a particular leaf, the proportions of different stomatal types are simply read off from the inherited curves.

(*e*) The more intimate short-range relationships between cells are settled upon the principles explained above (pp. 126–33).

This scheme is the outcome of a course of laboratory study which extended over several years and involved the reading of about 140 000 stomata. Normal statistical criteria were satisfied, and more limited enquiries in other groups of plants are indicative of similar behaviour-patterns elsewhere. It is inevitable that further work must lead to progressive correction and modification, but there is no way that any permanent harm could be done even if this initial statement of relationships turned out to be completely wrong; one man's mistake is another man's opportunity, and of such is the progress of science. Some of those to whom these observations have been shown have however objected to them on the more fundamental ground that such studies are outside the legitimate scope of biological enquiry. A scientific reply to this is neither necessary nor possible: the question seems to be one of individual repugnance in which the dominant components are of an ethical, aesthetic, or religious nature. The application of mathematical methods to organisms apparently still has the power to raise emotions which the application of chemical methods does not.

Envoi

So far as the general development of methods is concerned (as distinct from the description of particular tissue samples) we seem now to have reached that misty and uncomfortable region sometimes known as 'the frontiers of science'. It could perhaps be argued that our position is already somewhat too advanced, in that theoretical innovation has outstripped the productive capacity of our laboratories. Certainly a period of consolidation, of methodical application of the ideas now at our disposal, would be a prerequisite for any further large move in the theoretical field.

The author's personal contribution to this subject is probably now at an end, and is necessarily very imperfect. No doubt the geometrical analysis of tissues could be carried on in ways superior to anything which has been offered in these pages.

In general however it appears that the difficulties of the science have been absurdly exaggerated in the main stream of biological thought. The effect of mitosis in diversifying the shapes of cells is probably a much easier thing for a student to appreciate than, let us say, the genetical consequences of chiasmata, yet the shape of cells in the simplest tissues seems to be widely regarded as a mystery almost too deep for the mind of man to fathom. This might be excusable as an immediate response to an unfamiliar problem, but there is something seriously wrong with a science in which such an attitude can be maintained by so many for such a long time. As a *considered* intellectual position it seems undignified, and a little childish.

On the other hand it is necessary to question the easy and well-nigh universal assumption that tissue geometry has somehow been specially disconnected from the general body of biological science for the convenience of those who dislike mathematics. Where there is a tissue there will generally be a mitotic cycle, anisotropies, polarities, cell contacts, wave effects, and a system of idioblastic classification. There is no way of excluding these phenomena from the laboratory, and there is no way in which any aspect of tissue organisation (as distinct from matters pertaining only to a single cell) can fail to fall within one or other of the listed categories.

A histologist who says that he does not wish to concern himself with geometrical relationships is therefore in some measure abdicating from effective control of his enquiries. In formal terms he is stating a preference for imprecise objectives and incompletely standardised samples. Such a position would not be tenable for an instant in any of the more advanced sciences, but the biological community is as yet too uncritical and undemanding for the theoretical weaknesses of contemporary working methods to be seriously felt. Given a readership which asks for little more than loosely worded descriptive generalities with photographic illustrations, there is nothing to enforce absolute standards of theoretical discrimination upon a research worker. Nor are matters significantly altered when supplementary quantitative data are submitted to the judgement of persons who are, if not actually unable, at least quite unwilling, to write so much as a line or two of original algebra in the course of their assessment.

This book has therefore naturally not been directed primarily to those whose interests stop short at the descriptive level, though even the most passive contemplation of a histological section would seem likely, in time, to raise some question of geometry in the mind of the observer. The main purpose has been to offer encouragement (and perhaps even a little assistance) to all those who are disposed to take measurements, scribble equations, construct models and diagrams, or by any other means to extend the existing body of coherent quantitative thought upon the matters discussed. Obviously there are certain anomalous tissues (notably muscle) which must for the foreseeable future be separately treated as special cases, but after these reservations have been made there will remain an enormous field of potential application for the general methods which have been proposed. The value of those methods can be judged only by laboratory trial followed by personal evaluation of the 'interest' of the results obtained. The author can only say that whenever he has attempted any point of geometrical analysis upon an actual specimen he has obtained results which appeared to be biologically informative, and that the ratio of reward to effort has probably been about the same as one would hope to achieve in a genetical or taxonomic investigation. Our subject seems therefore to be quite capable of taking a more or less equal place alongside the various other styles of biological investigation. Whether it is really to attain such a position is a matter quite out of the hands of any individual. If geometrical analysis should prove more attractive to the quiet seeker after truth than to the angry controversialist, that perhaps may be no bad thing.

References

Bailey, I. W. & Shepard, H. B. (1915). Sanio's laws for variation in size of coniferous tracheids. *Bot. Gaz.* **60**, 66–71.

Bannan, M. W. (1966). Spiral grain and anticlinal divisions in the cambium of conifers. *Can. J. Bot.* **44**, 1515–38.

Bannan, M. W. (1968). Polarity in the survival and elongation of fusiform initials in conifer cambium. *Can. J. Bot.* **46**, 1005–8.

Bonner, J. T. (1961). Preface to Thompson (1961).

Dobell, C. (1949). D'Arcy Wentworth Thompson 1860–1948. *R. Soc. London Obituary Notices of Fellows* **6**, 599–617.

Dodd, J. D. (1944). Three-dimensional cell shape in the carpel vesicles of *Citrus grandis. Am. J. Bot.* **31**, 120–7.

Dodd, J. D. (1955). An approximation of the minimal tetrakaidekahedron. *Am. J. Bot.* **42**, 566–9.

Duffy, R. M. (1951). Comparative cellular configurations in the meristematic and mature cortical cells of the primary root of tomato. *Am. J. Bot.* **38**, 393–408.

Giesenhagen, K. (1909). Die Richtung der Teilungswand in Pflanzenzellen. *Flora, Jena* **99**, 355–69.

Glaser, O. & Child, G. P. (1937). The hexoctahedron and growth. *Biol Bull.* **73**, 205–13.

Graustein, W. C. (1931). On the average number of sides of polygons of a net. *Ann. Math.* series 2, **32**, 149–53.

Gross, P. L. K. (1927). The stackability of tetrakaidecahedra. *Science*, **66**, 131–2.

Harper, R. A. (1917). The evolution of cell types and contact and pressure responses in *Pediastrum. Mem. Torrey bot. Club*, **17**, 210–40.

Harper, R. A. (1918). Organization, reproduction, and inheritance in *Pediastrum. Proc. Am. Phil. Soc.* **57**, 375–439.

Hein, I. (1930). The tetrakaidecahedron in pseudoparenchyma. *Bull. Torrey bot. Club*, **57**, 59–62.

Hejnowicz, Z. (1973). Morphogenetic waves in cambia of trees. *Plant Sci. Lett.* **1**, 359–66.

Hejnowicz, Z. (1974). Pulsations of domain length as support for the hypothesis of morphogenetic waves in the cambium. *Acta Soc. Bot. Pol.* **43**, 261–71.

Holtzman, D. H. (1951). Three-dimensional cell shape studies in the vegetative tip of *Coleus. Am. J. Bot.* **38**, 221–34.

Hulbary, R. L. (1944). The influence of air spaces on the three-dimensional shapes of cells in Elodea stems, and a comparison with the pith cells of *Ailanthus. Am. J. Bot.* **31**, 561–80.

Hulbary, R. L. (1948). Three-dimensional cell shape in the tuberous roots of *Asparagus* and in the leaf of *Rhoeo. Am. J. Bot.* **35**, 558–66.

Ingold, C. T. (1973). Cell arrangement in coenobia of *Pediastrum. Ann. Bot. N.S.* **37**, 389–94.

Kelvin, Lord (as Sir William Thomson) (1887). On the division of space with minimal partitional area. *Phil. Mag.* 5s, **24**, 503–14.

Kelvin, Lord (1894). On the homogeneous division of space. *Proc. R. Soc. London*, **55**, 1–16.

Krawczyszyn, J. (1971). Unidirectional splitting and uniting of rays in the cambium of *Platanus* accompanying the formation of interlocked grain in wood. *Acta Soc. Bot. Pol.* **40**, 57–79.

Krawczyszyn, J. (1972). Movement of the cambial domain pattern and mechanism of formation of interlocked grain in *Platanus*. *Acta Soc. Bot. Pol.* **41**, 443–61.

Lewis, F. T. (1923). The typical shape of polyhedral cells in vegetable parenchyma and the restoration of that shape following cell division. *Proc. Am. Acad. Arts. Sci.* **58**, 537–52.

Lewis, F. T. (1925). A further study of the polyhedral shapes of cells. I. The stellate cells of *Juncus effusus*. II. Cells of human adipose tissue. III. Stratified cells of human oral epithelium. *Proc. Am. Acad. Arts Sci.* **61**, 1–34.

Lewis, F. T. (1926). The effect of cell division on the shape and size of hexagonal cells. *Anat. Rec.* **33**, 331–55.

Lewis, F. T. (1928). The correlation between cell division and the shapes and sizes of prismatic cells in the epidermis of *Cucumis*. *Anat. Rec.* **38**, 341–76.

Lewis, F. T. (1930). A volumetric study of growth and cell division in two types of epithelium – the longitudinally prismatic epidermal cells of *Tradescantia* and the radially prismatic epidermal cells of *Cucumis*. *Anat. Rec.* **47**, 59–99.

Lewis, F. T. (1931). A comparison between the mosaic of polygons in a film of artificial emulsion and the pattern of simple epithelium in surface view (cucumber epidermis and human amnion). *Anat. Rec.* **50**, 235–65.

Lewis, F. T. (1933a). Mathematically precise features of epithelial mosaics: observations on the endothelium of capillaries. *Anat. Rec.* **55**, 323–41.

Lewis, F. T. (1933b). The significance of cells as revealed by their polyhedral shapes, with special reference to precartilage, and a surmise concerning nerve cells and neuroglia. *Proc. Am. Acad. Arts Sci.* **68**, 251–84.

Lewis, F. T. (1935). The shape of the tracheids in the pine. *Am. J. Bot.* **22**, 741–62.

Lewis, F. T. (1943a). A geometric accounting for diverse shapes of 14-hedral cells: the transition from dodecahedra to tetrakaidecahedra. *Am. J. Bot.* **30**, 74–81.

Lewis, F. T. (1943b). The geometry of growth and cell division in epithelial mosaics. *Am. J. Bot.* **30**, 766–76.

Lewis, F. T. (1944). The geometry of growth and cell division in columnar parenchyma. *Am. J. Bot.* **31**, 619–29.

Lewis, F. T. (1949a). The analogous shapes of cells and bubbles. *Proc. Am. Acad. Arts Sci.* **77**, 147–86.

Lewis, F. T. (1949b). The correlation in size and shape between epidermal and subepidermal cells. *Proc. Natl. Acad. Sci., USA*, **35**, 506–12.

Lewis, F. T. (1950). Reciprocal cell division in epidermal and subepidermal cells. *Am. J. Bot.* **37**, 715–21.

Lier, F. G. (1952). A comparison of the three-dimensional shapes of cork cambium and cork cells in the stem of *Pelargonium hortorum* Bailey. *Bull. Torrey bot. Club*, **79**, 312–28, 371–92.

Macior, L. W. (1960). The tetrakaidecahedron and related cell forms in undifferentiated plant tissues. *Bull. Torrey bot. Club*, **87**, 99–138.

Macior, L. W. & Matzke, E. B. (1951). An experimental analysis of cell wall curvatures and approximation to minimal tetrakaidecahedra in the leaf parenchyma of *Rhoeo discolor*. *Am. J. Bot.* **38**, 783–93.

Marvin, J. W. (1936). The aggregation of orthic tetrakaidecahedra. *Science*, **83**, 188.

Marvin, J. W. (1939*a*). The shape of compressed lead shot and its relation to cell shape. *Am. J. Bot.* **26**, 280–8.

Marvin, J. W. (1939*b*). Cell shape studies in the pith of *Eupatorium purpureum*. *Am. J. Bot.* **26**, 487–504.

Marvin, J. W. (1944). Cell shape and cell volume relations in the pith of *Eupatorium perfoliatum* L. *Am. J. Bot.* **31**, 208–18.

Matzke, E. B. (1939). Volume–shape relationships in lead shot and their bearing on cell shapes. *Am. J. Bot.* **26**, 288–95.

Matzke, E. B. (1946). The three-dimensional shape of bubbles in foam – An analysis of the role of surface forces in three-dimensional cell shape determination. *Am. J. Bot.* **33**, 58–80.

Matzke, E. B. (1948). The three-dimensional shape of epidermal cells of the apical meristem of *Anacharis densa* (Elodea). *Am. J. Bot.* **35**, 323–32.

Matzke, E. B. (1949). Three-dimensional shape changes during cell division in the epidermis of the apical meristem of *Anacharis densa* (Elodea). *Am. J. Bot.* **36**, 584–95.

Matzke, E. B. & Duffy, R. M. (1955). The three-dimensional shape of interphase cells within the apical meristem of *Anacharis densa* (Elodea). *Am. J. Bot.* **42**, 937–45.

Matzke, E. B. & Duffy, R. M. (1956). Progressive three-dimensional shape changes of dividing cells within the apical meristem of *Anacharis densa*. *Am. J. Bot.* **43**, 205–25.

Matzke, E. B. & Nestler, J. (1946). Volume–shape relationships in variant foams. A further study of the rôle of surface forces in three-dimensional cell shape determination. *Am. J. Bot.* **33**, 130–44.

Millis, J. (1918). A geometric basis for physical and organic phenomena. *Science*, **48**, 353–60.

Millis, J. (1926). The shape of cells in masses. *Science*, **64**, 225–6.

Mozingo, W. N. (1951). Changes in the three-dimensional shape during growth and division of living epidermal cells in the apical meristem of *Phleum pratense* roots. *Am. J. Bot.* **38**, 495–511.

Naum, Y. R. & Matzke, E. B. (1955). The interrelationship between orthic tetrakaidecahedra and rhombic dodecahedra in aggregation series. *Bull. Torrey bot. Club*, **82**, 480–5.

Preston, R. D. (1974). *The physical biology of plant cell walls*. Chapman & Hall, London. pp. xiv + 491.

Pyszynski, W. (1972). Downward movement of the domain pattern in *Aesculus* cambium producing wavy-grained xylem. *Acta Soc. Bot. Pol.* **41**, 511–17.

Rahn, J. E. (1956). A study of the three-dimensional configurations of sporogenous cells in the anthers of *Trillium erectum* L. with special reference to the geometrical aspects of cell shape. *Bull. Torrey bot. Club*, **83**, 355–76.

Schüepp, O. (1966). *Meristeme*. (*Experientia* Supplementum 11). Birkhaüser Verlag, Basel und Stuttgart. pp. 253.

Seigerman, N. (1951). Three-dimensional cell shape in coconut endosperm. *Am. J. Bot.* **38**, 811–23.

Thompson, D. W. (1915). Morphology and mathematics. *Trans. R. Soc. Edinburgh*, **50**, 857–95.

Thompson, D. W. (1961). *On growth and form*. An abridged edition, edited by J. T. Bonner. Cambridge University Press. pp. xiv + 345.

Thomson, R. B. & Hull, K. L. (1934). Physical laws and the cellular organization of plants. *Bot. Gaz.* **95**, 511–14.

Wheeler, G. E. (1955). The effects of cell division on three-dimensional cell shape in *Aloe barbadensis*. *Am. J. Bot.* **42**, 855–65.

Williams, W. M. & Smith, C. S. (1952). A study of grain shape in an aluminium alloy and other applications of stereoscopic microradiography. *J. Met.* **4**, 755–65.

Index

147